PROBLEM-SOLVING CASES IN MICROSOFT® ACCESS™ AND EXCEL®

PROBLEM-SOLVING CASES IN MICROSOFT® ACCESS™ AND EXCEL®

Ninth Annual Edition

Ellen F. Monk

Joseph A. Brady

Gerard S. Cook

COURSE TECHNOLOGY
CENGAGE Learning™

Australia • Brazil • Japan • Korea • Mexico • Singapore • Spain • United Kingdom • United States

COURSE TECHNOLOGY
CENGAGE Learning™

**Problem-Solving Cases in Microsoft®
Access™ and Excel®, Ninth Annual Edition**
Ellen F. Monk, Joseph A. Brady,
Gerard S. Cook

Publisher: Joe Sabatino

Senior Acquisitions Editor: Charles
McCormick, Jr.

Senior Product Manager: Kate Mason

Development Editor: Dan Seiter

Editorial Assistant: Courtney Bavaro

Marketing Director: Keri Witman

Marketing Manager: Adam Marsh

Senior Marketing Communications Manager:
Libby Shipp

Marketing Coordinator: Suellen Ruttkay

Content Project Management:
PreMediaGlobal

Media Editor: Chris Valentine

Senior Art Director: Stacy Jenkins Shirley

Cover Designer: Lou Ann Thesing

Cover Image: iStock Photo

Manufacturing Coordinator: Julio Esperas

Compositor: PreMediaGlobal

For product information and technology assistance, contact us at
Cengage Learning Customer & Sales Support, 1-800-354-9706.
For permission to use material from this text or product,
submit all requests online at **cengage.com/permissions**.
Further permissions questions can be e-mailed to
permissionrequest@cengage.com.

Some of the product names and company names used in this book have been used for identification purposes only and may be trademarks or registered trademarks of their respective manufacturers and sellers.

Library of Congress Control Number: 2011920238

ISBN-13: 978-1-111-82051-0

ISBN-10: 1-111-82051-1

Course Technology
20 Channel Center Street
Boston, MA 02210
USA

Screenshots for this book were created using Microsoft Access and Excel®, and were used with permission from Microsoft.

Microsoft and the Office logo are either registered trademarks or trademarks of Microsoft Corporation in the United States and/or other countries. Course Technology, a part of Cengage Learning, is an independent entity from the Microsoft Corporation, and not affiliated with Microsoft in any manner.

The programs in this book are for instructional purposes only. They have been tested with care, but are not guaranteed for any particular intent beyond educational purposes. The author and the publisher do not offer any warranties or representations, nor do they accept any liabilities with respect to the programs.

Course Technology, a part of Cengage Learning, reserves the right to revise this publication and make changes from time to time in its content without notice.

Cengage Learning is a leading provider of customized learning solutions with office locations around the globe, including Singapore, the United Kingdom, Australia, Mexico, Brazil, and Japan. Locate your local office at: **www.cengage.com/global**.

Cengage Learning products are represented in Canada by Nelson Education, Ltd.

To learn more about Course Technology, visit **www.cengage.com/ coursetechnology**.

Purchase any of our products at your local college store or at our preferred online store **www.cengagebrain.com**.

Printed in the United States of America
1 2 3 4 5 6 7 17 16 15 14 13 12 11

BRIEF CONTENTS

Part 3: Decision Support Cases Using the Excel Solver

Part 4: Decision Support Case Using Basic Excel Functionality

Part 5: Integration Cases Using Access and Excel

Part 6: Advanced Skills Using Excel

Part 7: Presentation Skills

For two decades, we have taught MIS courses at the University of Delaware. From the start, we wanted to use good computer-based case studies for the database and decision-support portions of our courses.

At first, we could not find a casebook that met our needs! This surprised us because we thought our requirements were not unreasonable. First, we wanted cases that asked students to think about real-world business situations. Second, we wanted cases that provided students with hands-on experience, using the kind of software that they had learned to use in their computer literacy courses—and that they would later use in business. Third, we wanted cases that would strengthen students' ability to analyze a problem, examine alternative solutions, and implement a solution using software. Undeterred by the lack of casebooks, we wrote our own, and Course Technology, part of Cengage Learning, published it.

This is the ninth casebook we have written for Course Technology. The cases are all new, and the tutorials are updated.

As with our prior casebooks, we include tutorials that prepare students for the cases, which are challenging but doable. Most of the cases are organized to help students think about the logic of each case's business problem and then about how to use the software to solve the business problem. The cases fit well in an undergraduate MIS course, an MBA Information Systems course, or a Computer Science course devoted to business-oriented programming.

BOOK ORGANIZATION

The book is organized into seven parts:

- Database cases using Access
- Decision support cases using the Excel Scenario Manager
- Decision support cases using the Excel Solver
- Decision support case using basic Excel functionality
- Integration cases using Access and Excel
- Advanced skills using Excel
- Presentation skills

Part 1 begins with two tutorials that prepare students for the Access case studies. Parts 2 and 3 each begin with a tutorial that prepares students for the Excel case studies. All four tutorials provide students with hands-on practice in using the software's more advanced features—the kind of support that other books about Access and Excel do not provide. Part 4 asks students to use Excel's basic functionality for decision support. Part 5 challenges students to use both Access and Excel to find a solution to a business problem. Part 6 is a set of short tutorials that teach the advanced skills students need to complete some of the Excel cases. Part 7 is a tutorial that hones students' skills in creating and delivering an oral presentation to business managers. The next sections explore these parts of the book in more depth.

Part 1: Database Cases Using Access
This section begins with two tutorials and then presents five case studies.

Tutorial A: Database Design
This tutorial helps students understand how to set up tables to create a database, without requiring students to learn formal analysis and design methods, such as data normalization.

Tutorial B: Microsoft Access Tutorial
The second tutorial teaches students the more advanced features of Access queries and reports—features that students will need to know to complete the cases.

Cases 1–5

Five database cases follow Tutorials A and B. The students must use the Access database in each case to create forms, queries, and reports that help management. The first case is an easier "warm-up" case. The next four cases require more effort to design the database and implement the results.

Part 2: Decision Support Cases Using Excel Scenario Manager

This section has one tutorial and two decision support cases that require the use of the Excel Scenario Manager.

Tutorial C: Building a Decision Support System in Excel

This section begins with a tutorial that uses Excel to explain decision support and fundamental concepts of spreadsheet design. The case emphasizes the use of Scenario Manager to organize the output of multiple "what-if" scenarios.

Cases 6–7

Students can perform these two cases with or without Scenario Manager, although it is nicely suited to both cases. In each case, students must use Excel to model two or more solutions to a problem. Students then use the model outputs to identify and document the preferred solution in a memorandum. The instructor might also require students to summarize their solutions in an oral presentation.

Part 3: Decision Support Cases Using the Excel Solver

This section has one tutorial and two decision support cases that require the use of Excel Solver.

Tutorial D: Building a Decision Support System Using Excel Solver

This section begins with a tutorial for using Excel Solver, a powerful decision support tool for solving optimization problems.

Cases 8–9

Once again, students use Excel and the Solver tool in each case to analyze alternatives and identify and document the preferred solution.

Part 4: Decision Support Case Using Basic Excel Functionality
Case 10

The book continues with a case that uses basic Excel functionality. (In other words, the case does not require Scenario Manager or the Solver.) Excel is used to test student analytical skills in "what-if" analyses.

Part 5: Integration Cases Using Access and Excel
Cases 11 and 12

These cases integrate Access and Excel. The cases show students how to share data among multiple software packages to solve problems. Most students will think Case 11 is more difficult than Case 12.

Part 6: Advanced Skills Using Excel

This part contains one tutorial that focuses on using advanced techniques in Excel.

Tutorial E: Guidance for Excel Cases

A number of cases in this book require the use of advanced techniques in Excel. For example, techniques for using data tables and pivot tables are explained in Tutorial E rather than in the cases themselves.

Part 7: Presentation Skills
Tutorial F: Giving an Oral Presentation

Each case includes an optional assignment that lets students practice making a presentation to management to summarize the results of their case analysis. This tutorial gives advice for creating oral presentations. It also includes technical information on charting, a technique that is useful in case analyses or as support for presentations. This tutorial will help students to organize their recommendations, to present their solutions both in words and graphics, and to answer questions from the audience. For larger classes, instructors may want to have students work in teams to create and deliver their presentations, which would model the team approach used by many corporations.

To view and access additional cases, instructors should consider using the "Hall of Fame," as described in the *Using the Cases* section below.

INDIVIDUAL CASE DESIGN

The format of the cases uses the following template:

- Each case begins with a *Preview* and an overview of the tasks.
- The next section, *Preparation*, tells students what they need to do or know to complete the case successfully. Again, the tutorials also prepare students for the cases.
- The third section, *Background*, provides the business context that frames the case. The background of each case models situations that require the kinds of thinking and analysis that students will need in the business world.
- The *Assignment* sections are generally organized to help students develop their analyses.
- The last section, *Deliverables*, lists the finished materials that students must hand in: printouts, a memorandum, a presentation, and files on disk. The list is similar to the deliverables that a business manager might demand.

USING THE CASES

We have successfully used cases like these in our undergraduate MIS courses. We usually begin the semester with Access database instruction. We assign the Access database tutorials and then a case to each student. Then, to teach students how to use the Excel decision support system, we do the same thing: we assign a tutorial and then a case.

Some instructors have asked for access to extra cases, especially in the second semester of a school year. For example, they assigned the integration case in the fall, and need another one for the spring. To meet this need, we have set up an online "Hall of Fame" that features some of our favorite cases from prior editions. These password-protected cases are available to instructors on the Cengage Learning Web site. Go to *www.cengage.com/coursetechnology* and search for this textbook by title, author, or ISBN. Note that the cases are in MS Office 2010 format.

TECHNICAL INFORMATION

This textbook was tested for quality assurance using the Windows 7 operating system, Microsoft Access 2010, and Microsoft Excel 2010.

Data Files and Solution Files

We have created "starter" data files for the Excel cases, so students need not spend time typing in the spreadsheet skeleton. Cases 11 and 12 also require students to load an Access database file. All these files are on the Cengage Learning Web site, which is available both to students and instructors. Instructors should go to *www.cengage.com/coursetechnology* and search for this textbook by title, author, or ISBN. Students will find the files at *www.CengageBrain.com.* You are granted a license to copy the data files to any computer or computer network used by people who have purchased this textbook.

Solutions to the material in the text are available to instructors at *www.cengage.com/coursetechnology.* Search for this textbook by title, author, or ISBN. The solutions are password protected.

Instructor's Manual

An Instructor's Manual is available to accompany this text and help instructors use it successfully. The Instructor's Manual contains tools and information such as a sample syllabus, teaching tips, and grading guidelines. Instructors should go to *www.cengage.com/coursetechnology* and search for this textbook by title, author, or ISBN. The Instructor's Manual is password protected.

ACKNOWLEDGEMENTS

We would like to give many thanks to the team at Cengage Learning, including our Development Editor, Dan Seiter; Senior Product Manager, Kate Hennessy Mason; and our Content Project Manager, Karunakaran Gunasekaran. As always, we acknowledge our students' diligent work.

PART 1

DATABASE CASES USING ACCESS

TUTORIAL

DATABASE DESIGN

This tutorial has three sections. The first section briefly reviews basic database terminology. The second section teaches database design. The third section features a practice database design problem.

REVIEW OF TERMINOLOGY

You will begin by reviewing some basic terms that will be used throughout this textbook. In Access, a **database** is a group of related objects that are saved in one file. An Access **object** can be a table, form, query, or report. You can identify an Access database file by its suffix, .accdb.

A **table** consists of data that is arrayed in rows and columns. A **row** of data is called a **record**. A **column** of data is called a **field**. Thus, a record is a set of related fields. The fields in a table should be related to one another in some way. For example, a company might want to keep its employee data together by creating a database table called Employee. That table would contain data fields about employees, such as their names and addresses. It would not have data fields about the company's customers; that data would go in a Customer table.

A field's values have a **data type** that is declared when a table is defined. Thus, when data is entered into the database, the software knows how to interpret each entry. Data types in Access include the following:

- Text for words
- Integer for whole numbers
- Double for numbers that have a decimal value
- Currency for numbers that should be treated as dollars and cents
- Yes/No for variables that have only two values (such as 1/0, on/off, yes/no, and true/false)
- Date/Time for variables that are dates or times

Each database table should have a **primary key** field, a field in which each record has a *unique* value. For example, in an Employee table, a field called SSN (for Social Security number) could serve as a primary key, because each record's SSN value would be different. Sometimes a table does not have a single field whose values are all different. In that case, two or more fields are combined into a **compound primary key**. The combination of the fields' values is unique.

Database tables should be logically related to one another. For example, suppose a company has an Employee table with fields for SSN, Name, Address, and Telephone Number. For payroll purposes, the company has an Hours Worked table with a field that summarizes Labor Hours for individual employees. The relationship between the Employee table and Hours Worked table needs to be established in the database so you can determine the number of hours worked by any employee. To create this relationship, you include the primary key field from the Employee table (SSN) as a field in the Hours Worked table. In the Hours Worked table, the SSN field is then called a **foreign key**.

In Access, data can be entered directly into a table or it can be entered into a form, which then inserts the data into a table. A **form** is a database object that is created from an existing table to make the process of entering data more user-friendly.

A **query** is the database equivalent of a question that is posed about data in a table (or tables). For example, suppose a manager wants to know the names of employees who have worked for the company for more than five years. A query could be designed so that it interrogates the Employee table to search for the information. The query would be run, and its output would answer the question.

Because a query may need to pull data from more than one table, queries can be designed to interrogate multiple tables at a time. In that case, the tables must be connected by a **join** operation, which links tables on the values in a field that they have in common. The common field acts as a "hinge" for the joined tables; when the query is run, the query generator treats the joined tables as one large table.

In Access, queries that answer a question are called select queries because they select relevant data from the database records. Queries also can be designed to change data in records, add a record to the end of a table, or delete entire records from a table. These queries are called **update**, **append**, and **delete** queries, respectively.

Access has a **report** generator that can be used to format a table's data or a query's output.

DATABASE DESIGN

Designing a database involves determining which tables belong in the database and creating the fields that belong in each table. This section begins with an introduction to key database design concepts, then discusses design rules you should use when building a database. First, the following key concepts are defined:

- Entities
- Relationships
- Attributes

Database Design Concepts

Computer scientists have highly formalized ways of documenting a database's logic, and learning their notations and mechanics can be time-consuming and difficult. In fact, doing so usually takes a good portion of a systems analysis and design course. This tutorial will teach you database design by emphasizing practical business knowledge, and the approach should enable you to design serviceable databases quickly. Your instructor may add more formal techniques.

A database models the logic of an organization's operation, so your first task is to understand the operation. You can talk to managers and workers, make your own observations, and look at business documents such as sales records. Your goal is to identify the business's "entities" (sometimes called *objects*). An **entity** is a thing or event that the database will contain. Every entity has characteristics, called **attributes**, and one or more **relationships** to other entities. Take a closer look.

Entities

As previously mentioned, an entity is a tangible thing or an event. The reason for identifying entities is that *an entity eventually becomes a table in the database*. Entities that are things are easy to identify. For example, consider a video store. The database for the video store would probably need to contain the names of DVDs and the names of customers who rent them, so you would have one entity named Video and another named Customer.

In contrast, entities that are events can be more difficult to identify, probably because they are more conceptual. However, events are real, and they are important. In the video store example, one event would be Video Rental and another event would be Hours Worked by employees.

In general, your analysis of an organization's operations is made easier when you realize that organizations usually have physical entities such as these:

- Employees
- Customers
- Inventory (products or services)
- Suppliers

Thus, the database for most organizations would have a table for each of these entities. Your analysis also can be made easier by knowing that organizations engage in transactions internally (within the company) and externally (with the outside world). Such transactions are explained in an introductory accounting course, but most people understand them from events that occur in daily life. Consider the following examples:

- Organizations generate revenue from sales or interest earned. Revenue-generating transactions include event entities called Sales and Interest.
- Organizations incur expenses from paying hourly employees and purchasing materials from suppliers. Hours Worked and Purchases are event entities in the databases of most organizations.

Thus, identifying entities is a matter of observing what happens in an organization. Your powers of observation are aided by knowing what entities exist in the databases of most organizations.

Relationships

As an analyst building a database, you should consider the relationship of each entity to the other entities you have identified. For example, a college database might contain entities for Student, Course, and Section to contain data about each. A relationship between Student and Section could be expressed as "Students enroll in sections."

An analyst also must consider the **cardinality** of any relationship. Cardinality can be one-to-one, one-to-many, or many-to-many:

- In a one-to-one relationship, one instance of the first entity is related to just one instance of the second entity.
- In a one-to-many relationship, one instance of the first entity is related to many instances of the second entity, but each instance of the second entity is related to only one instance of the first.
- In a many-to-many relationship, one instance of the first entity is related to many instances of the second entity, and one instance of the second entity is related to many instances of the first.

For a more concrete understanding of cardinality, consider again the college database with the Student, Course, and Section entities. The university catalog shows that a course such as Accounting 101 can have more than one section: 01, 02, 03, 04, and so on. Thus, you can observe the following relationships:

- The relationship between the entities Course and Section is one-to-many. Each course has many sections, but each section is associated with just one course.
- The relationship between Student and Section is many-to-many. Each student can be in more than one section, because each student can take more than one course. Also, each section has more than one student.

Thinking about relationships and their cardinalities may seem tedious to you. However, as you work through the cases in this text, you will see that this type of analysis can be valuable in designing databases. In the case of many-to-many relationships, you should determine the tables a given database needs; in the case of one-to-many relationships, you should decide which fields the tables need to share.

Attributes

An attribute is a characteristic of an entity. You identify attributes of an entity because *attributes become a table's fields*. If an entity can be thought of as a noun, an attribute can be considered an adjective that describes the noun. Continuing with the college database example, consider the Student entity. Students have names, so Last Name would be an attribute of the Student entity and therefore a field in the Student table. First Name would be an attribute as well. The Student entity would have an Address attribute as another field, along with Phone Number and other descriptive fields.

Sometimes it can be difficult to tell the difference between an attribute and an entity, but one good way is to ask whether more than one attribute is possible for each entity. If more than one instance is possible, but you do not know the number in advance, you are working with an entity. For example, assume that a student could have a maximum of two addresses—one for home and one for college. You could specify attributes Address 1 and Address 2. Next, consider that you might not know the number of student addresses in advance, meaning that all addresses have to be recorded. In that case, you would not know how many fields to set aside in the Student table for addresses. Therefore, you would need a separate Student Addresses table (entity) that would show any number of addresses for a given student.

DATABASE DESIGN RULES

As described previously, your first task in database design is to understand the logic of the business situation. Once you understand this logic, you are ready to build the database. To create a context for learning about database design, look at a hypothetical business operation and its database needs.

Example: The Talent Agency

Suppose you have been asked to build a database for a talent agency that books musical bands into night-clubs. The agent needs a database to keep track of the agency's transactions and to answer day-to-day questions. For example, a club manager often wants to know which bands are available on a certain date at a certain time, or wants to know the agent's fee for a certain band. Similarly, the agent may want to see a list of all band members and the instrument each person plays, or a list of all bands that have three members.

Suppose that you have talked to the agent and have observed the agency's business operation. You conclude that your database needs to reflect the following facts:

1. A booking is an event in which a certain band plays in a particular club on a particular date, starting and ending at certain times, and performing for a specific fee. A band can play more than once a day. The Heartbreakers, for example, could play at the East End Cafe in the afternoon and then at the West End Cafe on the same night. For each booking, the club pays the talent agent. The agent keeps a five percent fee and then gives the remainder of the payment to the band.

2. Each band has at least two members and an unlimited maximum number of members. The agent notes a telephone number of just one band member, which is used as the band's contact number. No two bands have the same name or telephone number.

3. No members of any of the bands have the same name. For example, if one band has a member named Sally Smith, there is no Sally Smith in another band.

4. The agent keeps track of just one instrument that each band member plays. For the purpose of this database, "vocals" are considered an instrument.

5. Each band has a desired fee. For example, the Lightmetal band might want $700 per booking, and would expect the agent to try to get at least that amount.

6. Each nightclub has a name, an address, and a contact person. The contact person has a telephone number that the agent uses to call the club. No two clubs have the same name, contact person, or telephone number. Each club has a target fee. The contact person will try to get the agent to accept that fee for a band's appearance.

7. Some clubs feed the band members for free; others do not.

Before continuing with this tutorial, you might try to design the agency's database on your own. Ask yourself: What are the entities? Recall that business databases usually have Customer, Employee, and Inventory entities, as well as an entity for the event that generates revenue transactions. Each entity becomes a table in the database. What are the relationships between entities? For each entity, what are its attributes? For each table, what is the primary key?

Six Database Design Rules

Assume that you have gathered information about the business situation in the talent agency example. Now you want to identify the tables required for the database and the fields needed in each table. Observe the following six rules:

Rule 1: You do not need a table for the business. The database represents the entire business. Thus, in the example, Agent and Agency are not entities.

Rule 2: Identify the entities in the business description. Look for typical things and events that will become tables in the database. In the talent agency example, you should be able to observe the following entities:

- *Things*: The product (inventory for sale) is Band. The customer is Club.
- *Events*: The revenue-generating transaction is Bookings.

You might ask yourself: Is there an Employee entity? Isn't Instrument an entity? Those issues will be discussed as the rules are explained.

Rule 3: Look for relationships between the entities. Look for one-to-many relationships between entities. The relationship between those entities must be established in the tables, using a foreign key. For details, see the following discussion in Rule 4 about the relationship between Band and Band Member.

Look for many-to-many relationships between entities. Each of these relationships requires a third entity that associates the two entities in the relationship. Recall the many-to-many relationship from the college database scenario that involved Student and Section entities. To display the enrollment of specific students in specific sections, a third table would be required. The mechanics of creating such a table are described in Rule 4 during the discussion of the relationship between Band and Club.

Rule 4: Look for attributes of each entity and designate a primary key. As previously mentioned, you should think of the entities in your database as nouns. You should then create a list of adjectives that describe those nouns. These adjectives are the attributes that will become the table's fields. After you have identified fields for each table, you should check to see whether a field has unique values. If such a field exists, designate it as the primary key field; otherwise, designate a compound primary key.

In the talent agency example, the attributes, or fields, of the Band entity are Band Name, Band Phone Number, and Desired Fee, as shown in Figure A-1. No two bands have the same names, so the primary key field can be Band Name. The data type of each field is shown.

BAND	
Field Name	**Data Type**
Band Name (primary key)	Text
Band Phone Number	Text
Desired Fee	Currency

FIGURE A-1　The Band table and its fields

Two Band records are shown in Figure A-2.

Band Name (primary key)	Band Phone Number	Desired Fee
Heartbreakers	981 831 1765	$800
Lightmetal	981 831 2000	$700

FIGURE A-2　Records in the Band table

If two bands might have the same name, Band Name would not be a good primary key, so a different unique identifier would be needed. Such situations are common. Most businesses have many types of inventory, and duplicate names are possible. The typical solution is to assign a number to each product to use as the primary key field. For example, a college could have more than one faculty member with the same name, so each faculty member could be assigned an employee identification number (EIN). Similarly, banks assign a personal identification number (PIN) for each depositor. Each automobile produced by a car manufacturer gets a unique Vehicle Identification Number (VIN). Most businesses assign a number to each sale, called an invoice number. (The next time you go to a grocery store, note the number on your receipt. It will be different from the number on the next customer's receipt.)

At this point, you might be wondering why Band Member would not be an attribute of Band. The answer is that, although you must record each band member, you do not know in advance how many members will be in each band. Therefore, you do not know how many fields to allocate to the Band table for members. Another way to think about band members is that they are the agency's employees, in effect. Databases for organizations usually have an Employee entity. Therefore, you should create a Band Member table with the attributes Member Name, Band Name, Instrument, and Phone. The table and its fields are shown in Figure A-3.

BAND MEMBER	
Field Name	**Data Type**
Member Name (primary key)	Text
Band Name (foreign key)	Text
Instrument	Text
Phone	Text

FIGURE A-3　The Band Member table and its fields

Note in Figure A-3 that the phone number is classified as a Text data type because the field values will not be used in an arithmetic computation. The benefit is that Text data type values take up fewer bytes than Numerical or Currency data type values; therefore, the file uses less storage space. You should also use the Text data type for number values such as zip codes, Social Security numbers, and so on.

Five records in the Band Member table are shown in Figure A-4.

Member Name (primary key)	Band Name	Instrument	Phone
Pete Goff	Heartbreakers	Guitar	981 444 1111
Joe Goff	Heartbreakers	Vocals	981 444 1234
Sue Smith	Heartbreakers	Keyboard	981 555 1199
Joe Jackson	Lightmetal	Sax	981 888 1654
Sue Hoopes	Lightmetal	Piano	981 888 1765

FIGURE A-4   Records in the Band Member table

You can include Instrument as a field in the Band Member table because the agent records only one instrument for each band member. Thus, you can use the instrument as a way to describe a band member, much like the phone number is part of the description. Member Name can be the primary key because you can assume that no two members in any band have the same name. Alternatively, Phone could be the primary key, assuming that no two members share a telephone. Or, you could assign an ID number to each band member, which would create a unique identifier for each musician the agency handled.

You might ask why Band Name is included in the Band Member table. The commonsense reason is that you did not include the Member Name in the Band table. You must relate bands and members somewhere, and the Band Member table is the place to do it.

To think about this relationship in another way, consider the cardinality of the relationship between Band and Band Member. It is a one-to-many relationship: one band has many members, but each member is in just one band. You establish such a relationship in the database by using the primary key field of one table as a foreign key in the other table. In Band Member, the foreign key Band Name is used to establish the relationship between the member and his or her band.

The attributes of the Club entity are Club Name, Address, Contact Name, Club Phone Number, Preferred Fee, and Feed Band?. The Club table can define the Club entity, as shown in Figure A-5.

CLUB	
Field Name	**Data Type**
Club Name (primary key)	Text
Address	Text
Contact Name	Text
Club Phone Number	Text
Preferred Fee	Currency
Feed Band?	Yes/No

FIGURE A-5   The Club table and its fields

Two records in the Club table are shown in Figure A-6.

Club Name (primary key)	Address	Contact Name	Club Phone Number	Preferred Fee	Feed Band?
East End	1 Duce St.	Al Pots	981 444 8877	$600	Yes
West End	99 Duce St.	Val Dots	981 555 0011	$650	No

FIGURE A-6   Records in the Club table

You might wonder why Bands Booked into Club (or a similar name) is not an attribute of the Club table. There are two reasons. First, you do not know in advance how many bookings a club will have, so the value cannot be an attribute. Second, Bookings is the agency's revenue-generating transaction, an event entity, and you need a table for that business transaction. Consider the booking transaction next.

You know that the talent agent books a certain band into a certain club for a specific fee on a certain date, starting and ending at a specific time. From that information, you can see that the attributes of the Bookings entity are Band Name, Club Name, Date, Start Time, End Time, and Fee. The Bookings table and its fields are shown in Figure A-7.

BOOKINGS	
Field Name	**Data Type**
Band Name	Text
Club Name	Text
Date	Date/Time
Start Time	Date/Time
End Time	Date/Time
Fee	Currency

FIGURE A-7   The Bookings table and its fields—and no designation of a primary key

Some records in the Bookings table are shown in Figure A-8.

Band Name	Club Name	Date	Start Time	End Time	Fee
Heartbreakers	East End	11/21/10	21:30	23:30	$800
Heartbreakers	East End	11/22/10	21:00	23:30	$750
Heartbreakers	West End	11/28/10	19:00	21:00	$500
Lightmetal	East End	11/21/10	18:00	20:00	$700
Lightmetal	West End	11/22/10	19:00	21:00	$750

FIGURE A-8   Records in the Bookings table

Note that no single field is guaranteed to have unique values, because each band is likely to be booked many times and each club might be used many times. Furthermore, each date and time can appear more than once. Thus, no one field can be the primary key.

If a table does not have a single primary key field, you can make a compound primary key whose field values will be unique when taken together. Because a band can be in only one place at a time, one possible solution is to create a compound key from the Band Name, Date, and Start Time fields. An alternative solution is to create a compound primary key from the Club Name, Date, and Start Time fields.

To avoid having a compound key, you could create a field called Booking Number. Each booking would then have its own unique number, similar to an invoice number.

You can also think about this event entity in a different way. Over time, a band plays in many clubs, and each club hires many bands. Thus, Band and Club have a many-to-many relationship, which signals the need for a table between the two entities. A Bookings table would associate the Band and Club tables. You implement an associative table by including the primary keys from the two tables that are associated. In this case, the primary keys from the Band and Club tables are included as foreign keys in the Bookings table.

Rule 5: Avoid data redundancy. You should not include extra (redundant) fields in a table. Redundant fields take up extra disk space and lead to data entry errors because the same value must be entered in multiple tables, increasing the chance of a keystroke error. In large databases, keeping track of multiple instances of the same data is nearly impossible, so contradictory data entries become a problem.

Consider this example: Why wouldn't Club Phone Number be included in the Bookings table as a field? After all, the agent might have to call about a last-minute booking change and could quickly look up the number in the Bookings table. Assume that the Bookings table includes Booking Number as the primary key and Club Phone Number as a field. Figure A-9 shows the Bookings table with the additional field.

BOOKINGS	
Field Name	**Data Type**
Booking Number (primary key)	Text
Band Name	Text
Club Name	Text
Club Phone Number	Text
Date	Date/Time
Start Time	Date/Time
End Time	Date/Time
Fee	Currency

FIGURE A-9   The Bookings table with an unnecessary field—Club Phone Number

The fields Date, Start Time, End Time, and Fee logically depend on the Booking Number primary key— they help define the booking. Band Name and Club Name are foreign keys and are needed to establish the relationship between the Band, Club, and Bookings tables. But what about Club Phone Number? It is not defined by the Booking Number. It is defined by Club Name—*in other words, it is a function of the club, not of the booking*. Thus, the Club Phone Number field does not belong in the Bookings table. It is already in the Club table; if the agent needs the Club Phone Number field, it can be looked up there.

Perhaps you can see the practical data-entry problem of including Club Phone Number in Bookings. Suppose a club changed its contact phone number. The agent could easily change the number one time, in Club. But now the agent would need to remember all of the other tables that contained the field and change the values there too. In a small database, this task might not be difficult, but in larger databases, having redundant fields in many tables makes such maintenance difficult, which means that redundant data is often incorrect.

You might object by saying, "What about all of those foreign keys? Aren't they redundant?" In a sense, they are. But they are needed to establish the relationship between one entity and another, as discussed previously.

Rule 6: Do not include a field if it can be calculated from other fields. A **calculated field** is made using the query generator. Thus, the agent's fee is not included in the Bookings table because it can be calculated by query (here, five percent multiplied by the booking fee).

PRACTICE DATABASE DESIGN PROBLEM

Imagine that your town library wants to keep track of its business in a database, and that you have been called in to build the database. You talk to the town librarian, review the old paper-based records, and watch people use the library for a few days. You learn the following about the library:

1. Any resident of the town can get a library card simply by asking for one. The library considers each cardholder a member of the library.

2. The librarian wants to be able to contact members by telephone and by mail. She calls members when their books are overdue or when requested materials become available. She likes to mail a thank-you note to each patron on his or her anniversary of becoming a member of the library. Without a database, contacting members efficiently can be difficult; for example, multiple members can have the same name. Also, a parent and a child might have the same first and last name, live at the same address, and share a phone.

3. The librarian tries to keep track of each member's reading interests. When new books come in, the librarian alerts members whose interests match those books. For example, long-time member Sue Doaks is interested in reading Western novels, growing orchids, and baking bread. There must be some way to match her interests with available books. One complication is that, although the librarian wants to track all of a member's reading interests, she wants to classify each book as being in just one category of interest. For example, the classic gardening book *Orchids of France* would be classified as a book about orchids or a book about France, but not both.

4. The library stocks thousands of books. Each book has a title and any number of authors. Also, more than one book in the library might have the same title. Similarly, multiple authors might have the same name.

5. A writer could be the author of more than one book.

6. A book will be checked out repeatedly as time goes on. For example, *Orchids of France* could be checked out by one member in March, by another member in July, and by another member in September.

7. The library must be able to identify whether a book is checked out.

8. A member can check out any number of books in one visit. Also, a member might visit the library more than once a day to check out books.

9. All books that are checked out are due back in two weeks, with no exceptions. The late fee is 50 cents per day. The librarian would like to have an automated way of generating an overdue book list each day so she can telephone offending members.

10. The library has a number of employees. Each employee has a job title. The librarian is paid a salary, but other employees are paid by the hour. Employees clock in and out each day. Assume that all employees work only one shift per day and that all are paid weekly. Pay is deposited directly into an employee's checking account—no checks are hand-delivered. The database needs to include the librarian and all other employees.

Design the library's database, following the rules set forth in this tutorial. Your instructor will specify the format of your work. Here are a few hints in the form of questions:

- A book can have more than one author. An author can write more than one book. How would you describe the relationship between books and authors?

- The library lends books for free, of course. If you were to think of checking out a book as a sales transaction for zero revenue, how would you handle the library's revenue-generating event?

- A member can borrow any number of books at one checkout. A book can be checked out more than once. How would you describe the relationship between checkouts and books?

TUTORIAL **B**

MICROSOFT ACCESS TUTORIAL

Microsoft Access is a relational database package that runs on the Microsoft Windows operating system. This tutorial was prepared using Access 2010.

Before using this tutorial, you should know the fundamentals of Microsoft Access and know how to use Windows. This tutorial explains advanced Access skills you will need to complete database case studies. The tutorial concludes with a discussion of common Access problems and how to solve them.

As a precaution, always observe proper file-saving and closing procedures. To exit Access, click the File tab and select Close Database, then click the File tab and select Exit. You can also simply select the Exit option to return to Windows. Always end your work with these steps. If you remove your USB key or other portable storage device when database forms and tables are shown on the screen, you will lose your work.

To begin this tutorial, you will create a new database called Employee.

AT THE KEYBOARD

Open a new database. Click the File tab, select New from the menu, and then click Blank database from the Available Templates list. Name the database Employee. Click the file folder next to the filename to browse for the folder where you want to save the file. Otherwise, your file will be saved automatically in the Documents folder. Click the Create button.

Your opening screen should resemble the screen shown in Figure B-1.

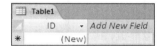

FIGURE B-1   Entering data in Datasheet view

When you create a table, Access opens it in Datasheet view by default. Because you will use Design view to build your tables, close the new table by clicking the *X* in the upper-right corner of the table window that corresponds to Close Table I. You are now on the Home tab in the Database window of Access, as shown in Figure B-2. From this screen, you can create or change objects.

FIGURE B-2   The Database window Home tab in Access

CREATING TABLES

Your database will contain data about employees, their wage rates, and the hours they worked.

Defining Tables

In the Database window, build three new tables using the following instructions.

AT THE KEYBOARD

Defining the Employee Table

This table contains permanent data about employees. To create the table, click the Create tab and then click Table Design in the Tables group. The table's fields are Last Name, First Name, Employee ID, Street Address, City, State, Zip, Date Hired, and US Citizen. The Employee ID field is the primary key field. Change the lengths of text fields from the default 255 spaces to more appropriate lengths; for example, the Last Name field might be 30 spaces, and the Zip field might be 10 spaces. Your completed definition should resemble the one shown in Figure B-3.

Field Name	Data Type	Description
Last Name	Text	
First Name	Text	
Employee ID	Text	
Street Address	Text	
City	Text	
State	Text	
Zip	Text	
Date Hired	Date/Time	
US Citizen	Yes/No	

FIGURE B-3   Fields in the Employee table

When you finish, click the File tab, select Save Object As, and then enter a name for the table. In this example, the table is named Employee. Make sure to specify the name of the *table*, not the database itself. (In this example, it is a coincidence that the Employee table has the same name as its database file.) Close the table by clicking the Close button (X) that corresponds to the Employee table.

Defining the Wage Data Table

This table contains permanent data about employees and their wage rates. The table's fields are Employee ID, Wage Rate, and Salaried. The Employee ID field is the primary key field. Use the data types shown in Figure B-4. Your definition should resemble the one shown in Figure B-4.

Field Name	Data Type	Description
Employee ID	Text	
Wage Rate	Currency	
Salaried	Yes/No	

FIGURE B-4   Fields in the Wage Data table

Click the File tab and then select Save Object As to save the table definition. Name the table Wage Data.

Defining the Hours Worked Table

The purpose of this table is to record the number of hours that employees work each week during the year. The table's three fields are Employee ID (which has a text data type), Week # (number–long integer), and Hours (number–double). The Employee ID and Week # are the compound keys.

In the following example, the employee with ID number 08965 worked 40 hours in Week 1 of the year and 52 hours in Week 2.

Employee ID	Week#	Hours
08965	1	40
08965	2	52

Note that no single field can be the primary key field because 08965 is an entry for each week. In other words, if this employee works each week of the year, 52 records will have the same Employee ID value at the end of the year. Thus, Employee ID values will not distinguish records. No other single field can distinguish these records either, because other employees will have worked during the same week number and some employees will have worked the same number of hours. For example, 40 hours—which corresponds to a full-time workweek—would be a common entry for many weeks.

All of this presents a problem because a table must have a primary key field in Access. The solution is to use a compound primary key; that is, use values from more than one field to create a combined field that will distinguish records. The best compound key to use for the current example consists of the Employee ID field and the Week # field, because as each person works each week, the week number changes. In other words, there is only *one* combination of Employee ID 08965 and Week # 1. Because those values *can occur in only one record*, the combination distinguishes that record from all others.

The first step of setting a compound key is to highlight the fields in the key. Those fields must appear one after the other in the table definition screen. (Plan ahead for that format.) As an alternative, you can highlight one field, hold down the Control key, and highlight the next field.

AT THE KEYBOARD

In the Hours Worked table, click the first field's left prefix area (known as the row selector), hold down the mouse button, and drag down to highlight the names of all fields in the compound primary key. Your screen should resemble the one shown in Figure B-5.

Field Name	Data Type	Description
Employee ID	Text	
Week #	Number	
Hours	Number	

FIGURE B-5   Selecting fields for the compound primary key for the Hours Worked table

Now click the Key icon. Your screen should resemble the one shown in Figure B-6.

Field Name	Data Type	Description
Employee ID	Text	
Week #	Number	
Hours	Number	

FIGURE B-6   The compound primary key for the Hours Worked table

You have created the compound primary key and finished defining the table. Click the File tab and then select Save Object As to save the table as Hours Worked.

Adding Records to a Table

At this point, you have set up the skeletons of three tables. The tables have no data records yet. If you printed the tables now, you would only see column headings (the field names). The most direct way to enter data into a table is to double-click the table's name in the navigation pane at the left side of the screen and then type the data directly into the cells.

NOTE

To display and open the database objects, Access 2010 uses a navigation pane, which is on the left side of the Access window.

AT THE KEYBOARD

On the Home tab of the Database window, double-click the Employee table. Your data entry screen should resemble the one shown in Figure B-7.

Employee										
Last Name ▾	First Name ▾	Employee ID ▾	Street Address ▾	City ▾	State ▾	Zip ▾	Date Hired ▾	US Citizen ▾	Add New Fi
*								☐	

FIGURE B-7   The data entry screen for the Employee table

The Employee table has many fields, some of which may be off the screen to the right. Scroll to see obscured fields. (Scrolling happens automatically as data is entered.) Figure B-7 shows all of the fields on the screen.

Enter your data one field value at a time. Note that the first row is empty when you begin. Each time you finish entering a value, press Enter to move the cursor to the next cell. After you enter data in the last cell in a row, the cursor moves to the first cell of the next row *and* Access automatically saves the record. Thus, you do not need to click the File tab and then select Save Object As after entering data into a table.

When entering data in your table, you should enter dates in the following format: 6/15/10. Access automatically expands the entry to the proper format in output.

Also note that Yes/No variables are clicked (checked) for Yes; otherwise, the box is left blank for No. You can change the box from Yes to No by clicking it.

Enter the data shown in Figure B-8 into the Employee table. If you make errors in data entry, click the cell, backspace over the error, and type the correction.

Employee								
Last Name ▾	First Name ▾	Employee ID ▾	Street Address ▾	City ▾	State ▾	Zip ▾	Date Hired ▾	US Citizen ▾
Howard	Jane	11411	28 Sally Dr	Glasgow	DE	19702	8/1/2010	☑
Smith	John	12345	30 Elm St	Newark	DE	19711	6/1/1996	☑
Smith	Albert	14890	44 Duce St	Odessa	DE	19722	7/15/1987	☑
Jones	Sue	22282	18 Spruce St	Newark	DE	19716	7/15/2004	☐
Ruth	Billy	71460	1 Tater Dr	Baltimore	MD	20111	8/15/1999	☐
Add	Your	Data	Here	Elkton	MD	21921		☑
*								☐

FIGURE B-8   Data for the Employee table

Note that the sixth record is *your* data record. Assume that you live in Elkton, Maryland, were hired on today's date (enter the date), and are a U.S. citizen. Make up a fictitious Employee ID number. For purposes of this tutorial, the sixth record has been created using the name of one of this text's authors and the employee ID 09911.

After adding records to the Employee table, open the Wage Data table and enter the data shown in Figure B-9.

Wage Data		
Employee ID ▾	Wage Rate ▾	Salaried ▾
11411	$10.00	☐
12345		☑
14890	$12.00	☐
22282		☑
71460		☑
Your Employee ID	$8.00	☐
*		☐

FIGURE B-9   Data for the Wage Data table

In this table, you are again asked to create a new entry. For this record, enter your own employee ID. Also assume that you earn $8 an hour and are not salaried. Note that when an employee's Salaried box is not checked (in other words, Salaried = No), the implication is that the employee is paid by the hour. Because salaried employees are not paid by the hour, their hourly rate is 0.00.

When you finish creating the Wage Data table, open the Hours Worked table and enter the data shown in Figure B-10.

Hours Worked		
Employee ID ▾	Week # ▾	Hours ▾
11411	1	40
11411	2	50
12345	1	40
12345	2	40
14890	1	38
14890	2	40
22282	1	40
22282	2	40
71460	1	40
71460	2	40
Your Employee ID	1	60
Your Employee ID	2	55

FIGURE B-10   Data for the Hours Worked table

Notice that salaried employees are always given 40 hours. Nonsalaried employees (including you) might work any number of hours. For your record, enter your fictitious employee ID, 60 hours worked for Week 1, and 55 hours worked for Week 2.

CREATING QUERIES

Because you know how to create basic queries, this section explains the advanced queries you will create in the cases in this book.

Using Calculated Fields in Queries

A **calculated field** is an output field made up of *other* field values. A calculated field is *not* a field in a table; it is created in the query generator. The calculated field does not become part of the table—it is just part of the query output. The best way to understand this process is to work through an example.

AT THE KEYBOARD

Suppose you want to see the employee IDs and wage rates of hourly workers, and the new wage rates if all employees were given a 10 percent raise. To view that information, show the employee ID, the current wage rate, and the higher rate, which should be titled New Rate in the output. Figure B-11 shows how to set up the query.

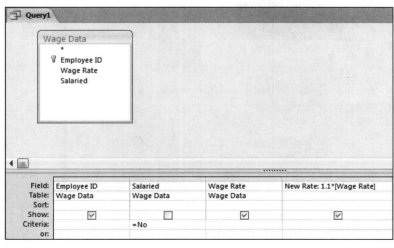

FIGURE B-11   Query setup for the calculated field

To set up this query, you need to select hourly workers by using the Salaried field with Criteria = No. Note in Figure B-11 that the Show box for the field is not checked, so the Salaried field values will not appear in the query output.

Note the expression for the calculated field, which you can see in the far-right field cell:

New Rate: 1.1 * [Wage Rate]

The term *New Rate:* merely specifies the desired output heading. (Don't forget the colon.) The rest of the expression, 1.1 * [Wage Rate], multiplies the old wage rate by 110 percent, which results in the 10 percent raise.

In the expression, the field name Wage Rate must be enclosed in square brackets. Remember this rule: *Any time an Access expression refers to a field name, the expression must be enclosed in square brackets*.

If you run this query, your output should resemble that in Figure B-12.

Query1		
Employee ID	Wage Rate	New Rate
11411	$10.00	11
14890	$12.00	13.2
09911	$8.00	8.8

FIGURE B-12   Output for a query with calculated field

Notice that the calculated field output is not shown in Currency format, but as a Double—a number with digits after the decimal point. To convert the output to Currency format, select the output column by clicking the line above the calculated field expression. The column darkens to indicate its selection. Your data entry screen should resemble the one shown in Figure B-13.

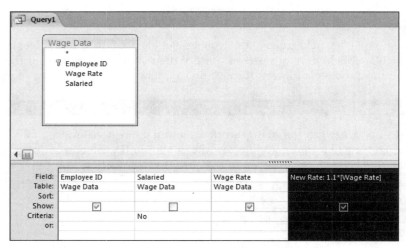

FIGURE B-13   Activating a calculated field in query design

Then, on the Design tab header, click Property Sheet in the Show/Hide group. The Field Properties window appears, as shown on the right in Figure B-14.

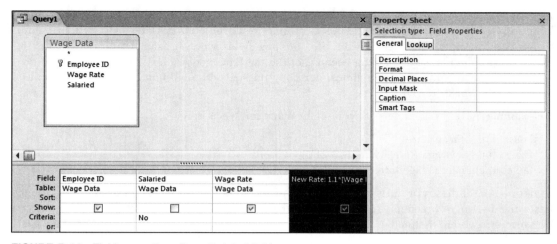

FIGURE B-14   Field properties of a calculated field

Click Format and choose Currency, as shown in Figure B-15. Then click the *X* in the upper-right corner of the window to close it.

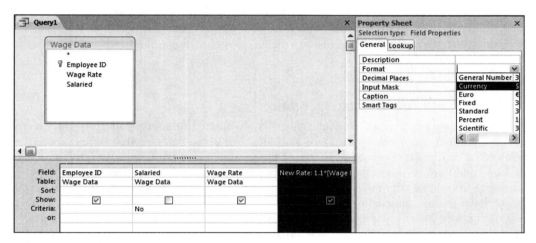

FIGURE B-15   Currency format of a calculated field

When you run the query, the output should resemble that in Figure B-16.

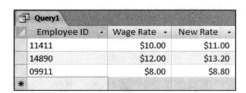

FIGURE B-16   Query output with formatted calculated field

Next, you examine how to avoid errors when making calculated fields.

Avoiding Errors when Making Calculated Fields

Follow these guidelines to avoid making errors in calculated fields:

- Do not enter the expression in the *Criteria* cell as if the field definition were a filter. You are making a field, so enter the expression in the *Field* cell.

- Spell, capitalize, and space a field's name *exactly* as you did in the table definition. If the table definition differs from what you type, Access thinks you are defining a new field by that name. Access then prompts you to enter values for the new field, which it calls a Parameter Query field. This problem is easy to debug because of the tag *Parameter Query*. If Access asks you to enter values for a parameter, you almost certainly misspelled a field name in an expression in a calculated field or criterion.

For example, here are some errors you might make for Wage Rate:

> Misspelling: (Wag Rate)
> Case change: (wage Rate / WAGE RATE)
> Spacing change: (WageRate / Wage Rate)

- Do not use parentheses or curly braces instead of the square brackets. Also, do not put parentheses inside square brackets. You *can*, however, use parentheses outside the square brackets in the normal algebraic manner.

For example, suppose that you want to multiply Hours by Wage Rate to get a field called Wages Owed. This is the correct expression:

> Wages Owed: [Wage Rate] * [Hours]

The following expression also would be correct:

> Wages Owed: ([Wage Rate] * [Hours])

But it would *not* be correct to omit the inside brackets, which is a common error:

> Wages Owed: [Wage Rate * Hours]

"Relating" Two or More Tables by the Join Operation

Often, the data you need for a query is in more than one table. To complete the query, you must **join** the tables by linking the common fields. One rule of thumb is that joins are made on fields that have common *values*, and those fields often can be key fields. The names of the join fields are irrelevant; also, the names of the tables or fields to be joined may be the same, but it is not required for an effective join.

Make a join by bringing in (adding) the tables needed. Next, decide which fields you will join. Then click one field name and hold down the left mouse button while you drag the cursor over to the other field's name in its window. Release the button. Access inserts a line to signify the join. (If a relationship between two tables has been formed elsewhere, Access inserts the line automatically, and you do not have to perform the click-and-drag operation. Access often inserts join lines without the user forming relationships.)

You can join more than two tables. The common fields *need not* be the same in all tables; that is, you can daisy-chain them together.

A common join error is to add a table to the query and then fail to link it to another table. In that case, you will have a table floating in the top part of the QBE (query by example) screen. When you run the query, your output will show the same records over and over. The error is unmistakable because there is *so much* redundant output. The two rules are to add only the tables you need and to link all tables.

Next, you will work through an example of a query that needs a join.

AT THE KEYBOARD

Suppose you want to see the last names, employee IDs, wage rates, salary status, and citizenship only for U.S. citizens and hourly workers. Because the data is spread across two tables, Employee and Wage Data, you should add both tables and pull down the five fields you need. Then you should add the Criteria expressions. Set up your work to resemble that in Figure B-17. Make sure the tables are joined on the common field, Employee ID.

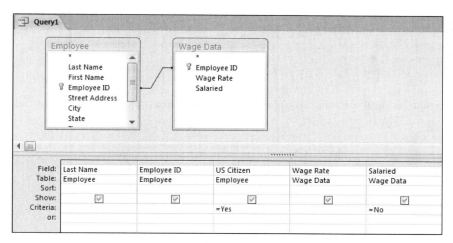

FIGURE B-17 A query based on two joined tables

You should quickly review the criteria you will need to set up this join: If you want data for employees who are U.S. citizens *and* who are hourly workers, the Criteria expressions go in the *same* Criteria row. If you want data for employees who are U.S. citizens *or* who are hourly workers, one of the expressions goes in the second Criteria row (the one with the or: notation).

Now run the query. The output should resemble that in Figure B-18, with the exception of the name "Brady."

Last Name	Employee ID	US Citizen	Wage Rate	Salaried
Howard	11411	☑	$10.00	☐
Smith	14890	☑	$12.00	☐
Brady	09911	☑	$8.00	☐
*		☐		☐

FIGURE B-18 Output of a query based on two joined tables

You do not need to print or save the query output, so return to Design view and close the query. Another practice query follows.

AT THE KEYBOARD

Suppose you want to see the wages owed to hourly employees for Week 2. You should show the last name, the employee ID, the salaried status, the week #, and the wages owed. Wages will have to be a calculated field ([Wage Rate] * [Hours]). The criteria are No for Salaried and 2 for the Week #. (This means that another "And" query is required.) Your query should be set up like the one in Figure B-19.

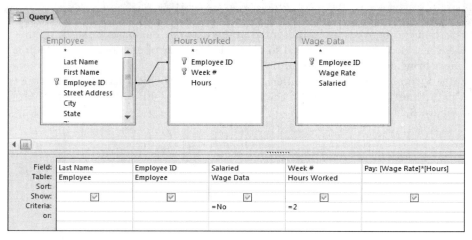

FIGURE B-19 Query setup for wages owed to hourly employees for Week 2

NOTE

In the query in Figure B-19, the calculated field column was widened so you could see the whole expression. To widen a column, click the column boundary line and drag to the right.

Run the query. The output should be similar to that in Figure B-20, if you formatted your calculated field to Currency.

Last Name	Employee ID	Salaried	Week #	Pay
Howard	11411	☐	2	$500.00
Smith	14890	☐	2	$480.00
Brady	09911	☐	2	$440.00
*		☐		

FIGURE B-20   Query output for wages owed to hourly employees for Week 2

Notice that it was not necessary to pull down the Wage Rate and Hours fields to make the query work. You do not need to save or print the query output, so return to Design view and close the query.

Summarizing Data from Multiple Records (Totals Queries)

You may want data that summarizes values from a field for several records (or possibly all records) in a table. For example, you might want to know the average hours that all employees worked in a week or the total (sum) of all of the hours worked. Furthermore, you might want data grouped or stratified in some way. For example, you might want to know the average hours worked, grouped by all U.S. citizens versus all non-U.S. citizens. Access calls such a query a **Totals query**. These queries include the following operations:

Sum	The total of a given field's values
Count	A count of the number of instances in a field—that is, the number of records. In the current example, you would count the number of employee IDs to get the number of employees.
Average	The average of a given field's values
Min	The minimum of a given field's values
Var	The variance of a given field's values
StDev	The standard deviation of a given field's values
Where	The field has criteria for the query output

AT THE KEYBOARD

Suppose you want to know how many employees are represented in the example database. First, bring the Employee table into the QBE screen. Because you will need to count the number of employee IDs, which is a Totals query operation, you must bring down the Employee ID field.

To tell Access that you want a Totals query, click the Design tab and then click the Totals icon in the Show/Hide group. A new row called the Total row opens in the lower part of the QBE screen. At this point, the screen resembles that in Figure B-21.

FIGURE B-21   Totals query setup

Note that the Total cell contains the words *Group By*. Until you specify a statistical operation, Access assumes that a field will be used for grouping (stratifying) data.

To count the number of employee IDs, click next to Group By to display an arrow. Click the arrow to reveal a drop-down menu, as shown in Figure B-22.

FIGURE B-22   Choices for statistical operation in a Totals query

Select the Count operator. (You might need to scroll down the menu to see the operator you want.) Your screen should resemble the one shown in Figure B-23.

FIGURE B-23   Count in a Totals query

Run the query. Your output should resemble that in Figure B-24.

FIGURE B-24   Output of Count in a Totals query

Notice that Access created a pseudo-heading, "CountOfEmployee ID," by splicing together the statistical operation (Count), the word Of, and the name of the field (Employee ID). If you wanted a phrase such as Count of Employees as a heading, you would go to Design view and change the query to resemble the one shown in Figure B-25.

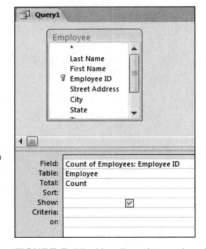

FIGURE B-25   Heading change in a Totals query

When you run the query, the output should resemble that in Figure B-26.

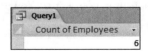

FIGURE B-26   Output of heading change in a Totals query

You do not need to print or save the query output, so return to Design view and close the query.

AT THE KEYBOARD

As another example of a Totals query, suppose you want to know the average wage rate of employees, grouped by whether the employees are salaried. Figure B-27 shows how to set up your query.

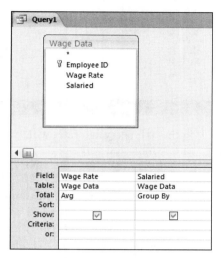

FIGURE B-27   Query setup for average wage rate of employees

When you run the query, your output should resemble that in Figure B-28.

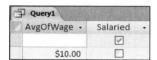

FIGURE B-28   Output of query for average wage rate of employees

Recall the convention that salaried workers are assigned zero dollars an hour. Suppose you want to eliminate the output line for zero dollars an hour because only hourly-rate workers matter for the query. The query setup is shown in Figure B-29.

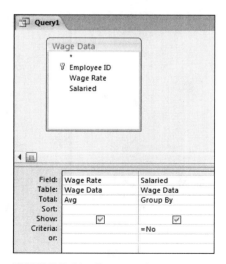

FIGURE B-29   Query setup for nonsalaried workers only

When you run the query, you will get output for nonsalaried employees only, as shown in Figure B-30.

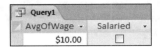

FIGURE B-30 Query output for nonsalaried workers only

Thus, it is possible to use Criteria in a Totals query, just as you would with a "regular" query. You do not need to print or save the query output, so return to Design view and close the query.

AT THE KEYBOARD

Assume that you want to see two pieces of information for hourly workers: (1) the average wage rate, which you will call Average Rate in the output; and (2) 110 percent of the average rate, which you will call the Increased Rate. To get this information, you can make a calculated field in a new query from a Totals query. In other words, you use one query as a basis for another query.

Create the first query; you already know how to perform certain tasks for this query. The revised heading for the average rate will be Average Rate, so type *Average Rate: Wage Rate* in the Field cell. Note that you want the average of this field. Also, the grouping will be by the Salaried field. (To get hourly workers only, enter *Criteria: No.*) Confirm that your query resembles that in Figure B-31, then save the query and close it.

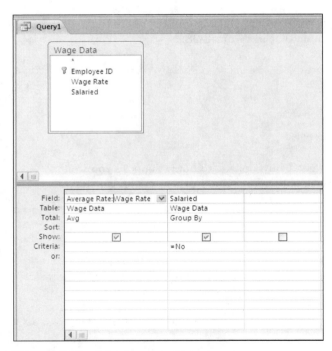

FIGURE B-31 A totals query with average

Now begin a new query. However, instead of bringing in a table to the query design, select a query. To start a new query, click the Create tab and then click the Query Design button in the Queries group. The Show Table dialog box appears. Click the Queries tab instead of using the default Tables tab, and select the query you just saved as a basis for the new query. The most difficult part of this query is to construct the expression for the calculated field. Conceptually, it is as follows:

Increased Rate: 1.1 * [The current average]

You use the new field name in the new query as the current average, and you treat the new name like a new field:

Increased Rate: 1.1 * [Average Rate]

The query within a query is shown in Figure B-32.

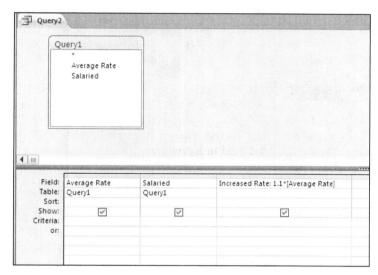

FIGURE B-32 A query within a query

Figure B-33 shows the output of the new query. Note that the calculated field is formatted.

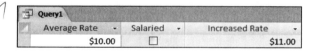

FIGURE B-33 Output of an Expression in a Totals query

You do not need to print or save the query output, so return to Design view and close the query.

Using the Date() Function in Queries

Access has two important date function features:

- The built-in Date() function gives you today's date. You can use the function in query criteria or in a calculated field. The function "returns" the day on which the query is run; in other words, it inserts the value where the Date() function appears in an expression.
- Date arithmetic lets you subtract one date from another to obtain the difference—in number of days—between two calendar dates. For example, suppose you create the following expression: 10/9/2009 – 10/4/2009
  Access would evaluate the expression as the integer 5 (9 minus 4 is 5).

As another example of how date arithmetic works, suppose you want to give each employee a one-dollar bonus for each day the employee has worked for you. You would need to calculate the number of days between the employee's date of hire and the day the query is run, and then multiply that number by $1.

You would find the number of elapsed days by using the following equation:

Date() – [Date Hired]

Also suppose that for each employee, you want to see the last name, employee ID, and bonus amount. You would set up the query as shown in Figure B-34.

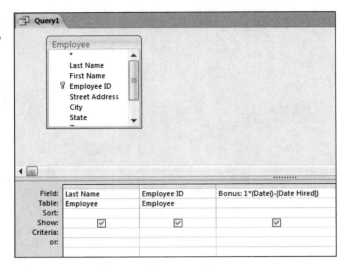

FIGURE B-34   Date arithmetic in a query

Assume that you set the format of the Bonus field to Currency. The output will be similar to that in Figure B-35, although your Bonus data will be different because you used a different date.

Last Name	Employee ID	Bonus
Brady	09911	$0.00
Howard	11411	$137.00
Smith	12345	$4,581.00
Smith	14890	$7,825.00
Jones	22282	$1,615.00
Ruth	71460	$3,411.00

FIGURE B-35   Output of query with date arithmetic

Using Time Arithmetic in Queries

Access also allows you to subtract the values of time fields to get an elapsed time. Assume that your database has a Job Assignments table showing the times that nonsalaried employees were at work during a day. The definition is shown in Figure B-36.

Table1

Field Name	Data Type
Employee ID	Text
ClockIn	Date/Time
ClockOut	Date/Time
DateWorked	Date/Time

FIGURE B-36   Date/Time data definition in the Job Assignments table

Assume that the DateWorked field is formatted for Long Date and that the ClockIn and ClockOut fields are formatted for Medium Time. Also assume that for a particular day, nonsalaried workers were scheduled as shown in Figure B-37.

Job Assignments

Employee ID	ClockIn	ClockOut	DateWorked	Click to Add
09911	8:30:00 AM	4:30:00 PM	Friday, September 30, 2011	
11411	9:00:00 AM	3:00:00 PM	Friday, September 30, 2011	
14890	7:00:00 AM	5:00:00 PM	Friday, September 30, 2011	

FIGURE B-37   Display of date and time in a table

q

You want a query showing the elapsed time that your employees were on the premises for the day. When you add the tables, your screen may show the links differently. Click and drag the Job Assignments, Employee, and Wage Data table icons to look like those in Figure B-38.

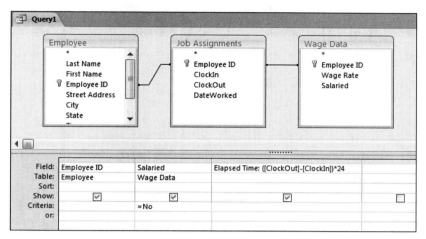

FIGURE B-38   Query setup for time arithmetic

Figure B-39 shows the output, which looks correct. For example, employee 09911 was at work from 8:30 a.m. to 4:30 p.m., which is eight hours. But how does the odd expression that follows yield the correct answers?

Query1		
Employee ID ▾	Salaried ▾	Elapsed Time ▾
09911	☐	8
11411	☐	6
14890	☐	10
*	☐	

-*not showing up ?. (mine)*

FIGURE B-39   Query output for time arithmetic

([ClockOut] – [ClockIn]) * 24

Why wouldn't the following expression work?

[ClockOut] – [ClockIn]

Here is the answer: In Access, subtracting one time from the other yields the *decimal* portion of a 24-hour day. Returning to the example, you can see that employee 09911 worked eight hours, which is one-third of a day, so the time arithmetic function yields .3333. That is why you must multiply by 24—to convert from decimals to an hourly basis. Hence, for employee 09911, the expression performs the following calculation: 1/3 × 24 = 8.

Note that parentheses are needed to force Access to do the subtraction *first*, before the multiplication. Without parentheses, multiplication takes precedence over subtraction. For example, consider the following expression:

[ClockOut] – [ClockIn] * 24

In this example, ClockIn would be multiplied by 24, the resulting value would be subtracted from Clock-Out, and the output would be a nonsensical decimal number.

Deleting and Updating Queries

The queries presented in this tutorial thus far have been Select queries. They select certain data from specific tables based on a given criterion. You also can create queries to update the original data in a database. Businesses use such queries often, and in real time. For example, when you order an item from a Web site, the company's database is updated to reflect your purchase through the deletion of that item from the company's inventory.

Consider an example. Suppose you want to give all nonsalaried workers a $0.50 per hour pay raise. Because you have only three nonsalaried workers, it would be easy to change the Wage Rate data in the table. However, if you had 3,000 nonsalaried employees, it would be much faster and more accurate to change the Wage Rate data by using an Update query that adds $0.50 to each nonsalaried employee's wage rate.

AT THE KEYBOARD

Now you will change each of the nonsalaried employees' pay via an Update query. Figure B-40 shows how to set up the query.

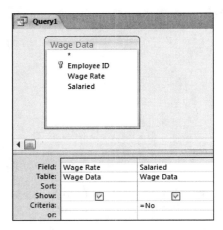

FIGURE B-40   Query setup for an Update query

So far, this query is just a Select query. Click the Update button in the Query Type group, as shown in Figure B-41.

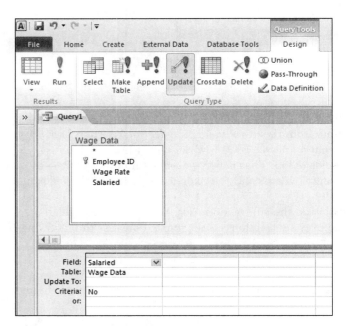

FIGURE B-41   Selecting a query type

Notice that you now have another line on the QBE grid called Update To:, which is where you specify the change or update to the data. Notice that you will update only the nonsalaried workers by using a filter under the Salaried field. Update the Wage Rate data to Wage Rate plus $0.50, as shown in Figure B-42. Note that the update involves the use of brackets [ ], as in a calculated field.

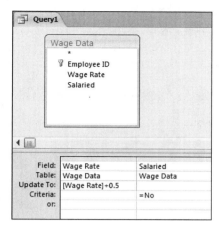

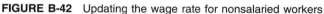

Expression you entered contains invalid syntax.

FIGURE B-42   Updating the wage rate for nonsalaried workers

Now run the query by clicking the Run button in the Results group. If you cannot run the query because it is blocked by Disabled Mode, click the Database Tools tab, then click Message Bar in the Show/Hide group. Click the Options button, choose "enable this content," and then click OK. When you successfully run the query, the warning message in Figure B-43 appears.

10

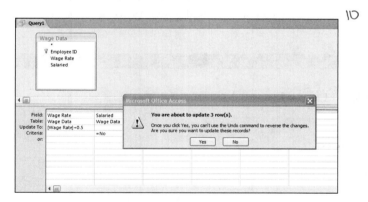

FIGURE B-43   Update query warning

When you click Yes, the records are updated. Check the updated records by viewing the Wage Data table. Each nonsalaried wage rate should be increased by $0.50. You could add or subtract data from another table as well. If you do, remember to put the field name in square brackets.

Another type of query is the Delete query, which works like Update queries. For example, assume that your company has been purchased by the state of Delaware, which has a policy of employing only state residents. Thus, you must delete (or fire) all employees who are not exclusively Delaware residents. To do that, you would create a Select query. Using the Employee table, you would click the Delete icon in the Query Type group, then bring down the State field and filter only those records that were not in Delaware (DE). Do not perform the operation, but note that if you did, the setup would look like the one in Figure B-44.

FIGURE B-44   Deleting all employees who are not Delaware residents

Using Parameter Queries

A **Parameter query** is actually a type of Select query. For example, suppose your company has 5,000 employees and you want to query the database to find the same kind of information again and again, but about different employees each time. For example, you might want to query the database to find out how many hours a particular employee has worked. You could run a query that you created and stored previously, but run it only for a particular employee.

AT THE KEYBOARD

Create a Select query with the format shown in Figure B-45.

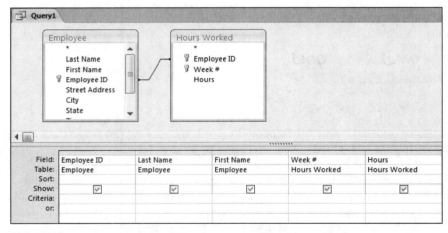

FIGURE B-45   Design of a Parameter query beginning as a Select query

In the Criteria line of the QBE grid for the Employee ID field, type what is shown in Figure B-46.

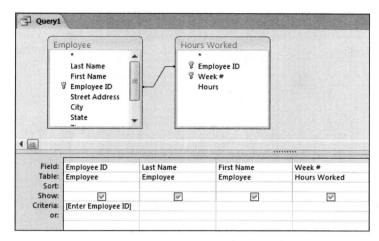

FIGURE B-46 Design of a Parameter query, continued

Note that the Criteria line uses square brackets, as you would expect to see in a calculated field. Now run the query. You will be prompted for the employee's ID number, as shown in Figure B-47.

FIGURE B-47 Enter Parameter Value dialog box

Enter your own employee ID. Your query output should resemble that in Figure B-48.

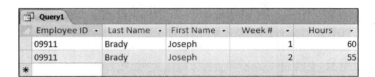

FIGURE B-48 Output of a Parameter query

MAKING SEVEN PRACTICE QUERIES

This portion of the tutorial gives you additional practice in creating queries. Before making these queries, you must create the specified tables and enter the records shown in the "Creating Tables" section of this tutorial. The output shown for the practice queries is based on those inputs.

AT THE KEYBOARD

For each query that follows, you are given a problem statement and a "scratch area." You also are shown what the query output should look like. Set up each query in Access and then run the query. When you are satisfied with the results, save the query and continue with the next one. Note that you will work with the Employee, Hours Worked, and Wage Data tables.

1. Create a query that shows the employee ID, last name, state, and date hired for employees who live in Delaware *and* were hired after 12/31/99. Perform an ascending sort by employee ID. First click the Sort cell of the field, and then choose Ascending or Descending. Before creating your query, use the table shown in Figure B-49 to work out your QBE grid on paper.

Field	Emp. ID	Last Name	State	Date Hired	
Table	Emp	Emp	Emp	Emp.	
Sort			Ascend.		
Show	✓	✓	✓	✓	
Criteria			"DE"	> 12/31/99	
Or:					

FIGURE B-49   QBE grid template

Your output should resemble that in Figure B-50.

Employee ID ▾	Last Name ▾	State ▾	Date Hired ▾
11411	Howard	DE	8/1/2010
22282	Jones	DE	7/15/2004

FIGURE B-50   Number 1 query output

2. Create a query that shows the last name, first name, date hired, and state for employees who live in Delaware *or* were hired after 12/31/99. The primary sort (ascending) is on last name, and the secondary sort (ascending) is on first name. The Primary Sort field must be to the left of the Secondary Sort field in the query setup. Before creating your query, use the table shown in Figure B-51 to work out your QBE grid on paper.

Field	Last Name	First Name	State	Date Hired	
Table	Emp.	Emp	Emp	Emp	
Sort	Ascend				
Show	✓	✓	✓	✓	
Criteria			"DE"		
Or:				> 12/31/99	

FIGURE B-51   QBE grid template

If your name was Joe Brady, your output would look like that in Figure B-52.

Last Name ▾	First Name ▾	Date Hired ▾	State ▾
Brady	Joseph	9/14/2011	MD
Howard	Jane	8/1/2010	DE
Jones	Sue	7/15/2004	DE
Smith	Albert	7/15/1987	DE
Smith	John	6/1/1996	DE

FIGURE B-52   Number 2 query output

3. Create a query that sums the number of hours worked by U.S. citizens and the number of hours worked by non-U.S. citizens. In other words, create two sums, grouped on citizenship. The heading for total hours worked should be Total Hours Worked. Before creating your query, use the table shown in Figure B-53 to work out your QBE grid on paper.

Field	Tot.Hrs: Hours	US Cit.			
Table	Hours Wkd	Emp.			
Total	Sum	Group By			
Sort		Ascend			
Show	✓	✓			
Criteria					
Or:					

FIGURE B-53 QBE grid template

Your output should resemble that in Figure B-54.

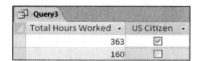

Total Hours Worked	US Citizen
363	☑
160	☐

FIGURE B-54 Number 3 query output

4. Create a query that shows the wages owed to hourly workers for Week 1. The heading for the wages owed should be Total Owed. The output headings should be Last Name, Employee ID, Week #, and Total Owed. Before creating your query, use the table shown in Figure B-55 to work out your QBE grid on paper.

Field	Last Name	Emp. ID	Week #	Tot.Owed: [Wage] [Hrs]	Salaried
Table	Emp.	Emp	Hrs Wkd		Wage Data
Sort					
Show	✓	✓	✓	✓	
Criteria			1		No
Or:					

FIGURE B-55 QBE grid template

If your name was Joe Brady, your output would look like that in Figure B-56.

Last Name	Employee ID	Week #	Total Owed
Howard	11411	1	$420.00
Smith	14890	1	$475.00
Brady	09911	1	$510.00

FIGURE B-56 Number 4 query output

5. Create a query that shows the last name, employee ID, hours worked, and overtime amount owed for hourly employees who earned overtime during Week 2. Overtime is paid at 1.5 times the normal hourly rate for all hours worked over 40. Note that the amount shown in the query should be just the overtime portion of the wages paid. Also, this is not a Totals query—amounts should be shown for individual workers. Before creating your query, use the table shown in Figure B-57 to work out your QBE grid on paper.

Field	last name	Emp ID	Hours	OT Pay 1.5*[WageRate]*([Hours]-40)	
Table	Emp	Emp	Hrs Wkd		
Sort					
Show	✓	✓	✓	✓	
Criteria			> 40	[Emp] 40	
Or:					

FIGURE B-57   QBE grid template

If your name was Joe Brady, your output would look like that in Figure B-58.

Last Name	Employee ID	Hours	OT Pay
Howard	11411	50	$157.50
Brady	09911	55	$191.25

FIGURE B-58   Number 5 query output

6. Create a Parameter query that shows the hours employees have worked. Have the Parameter query prompt for the week number. The output headings should be Last Name, First Name, Week #, and Hours. This query is for nonsalaried workers only. Before creating your query, use the table shown in Figure B-59 to work out your QBE grid on paper.

Field	last name	first name	week #	Hours	
Table	emp	emp	hrs wkd	hrs wkd	
Sort					
Show	✓	✓	✓	✓	
Criteria			[week #]		
Or:					

FIGURE B-59   QBE grid template

Run the query and enter 2 when prompted for the Week #. Your output should look like that in Figure B-60.

Last Name	First Name	Week #	Hours
Howard	Jane	2	50
Smith	Albert	2	40
Brady	Joseph	2	55

FIGURE B-60   Number 6 query output

7. Create an Update query that gives certain workers a merit raise. First, you must create an additional table, as shown in Figure B-61.

Merit Raises		
Employee ID ▾	Merit Raise ▾	Add New Field
11411	$0.25	
14890	$0.15	
*		

FIGURE B-61   Merit Raises table

8. Create a query that adds the Merit Raise to the current Wage Rate for employees who will receive a raise. When you run the query, you should be prompted with *You are about to update two rows*. Check the original Wage Data table to confirm the update. Before creating your query, use the table shown in Figure B-62 to work out your QBE grid on paper.

Field					
Table					
Update to					
Criteria					
Or:					

FIGURE B-62   QBE grid template

CREATING REPORTS

Database packages let you make attractive management reports from a table's records or from a query's output. If you are making a report from a table, the Access report generator looks up the data in the table and puts it into report format. If you are making a report from a query's output, Access runs the query in the background (you do not control it or see it happen) and then puts the output in report format.

There are different ways to make a report. One method is to hand-craft the report from scratch in Design view, but this tedious process is not explained in this tutorial. A simpler way is to select the query or table on which the report is based and then click Create Report. This streamlined method of creating reports is explained in this tutorial.

Creating a Grouped Report

This tutorial assumes that you already know how to create a basic ungrouped report, so this section teaches you how to make a grouped report. If you do not know how to create an ungrouped report, you can learn by following the first example in the upcoming section.

AT THE KEYBOARD

Suppose you want to create a report from the Hours Worked table. Select the table by clicking it once. Click the Create tab, then click Report in the Reports group. A report appears, as shown in Figure B-63.

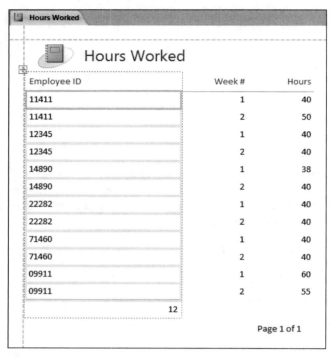

FIGURE B-63   Initial report based on a table

On the Design tab, select the Group and Sort button in the Grouping and Totals group. Your report will have an additional selection at the bottom, as shown in Figure B-64.

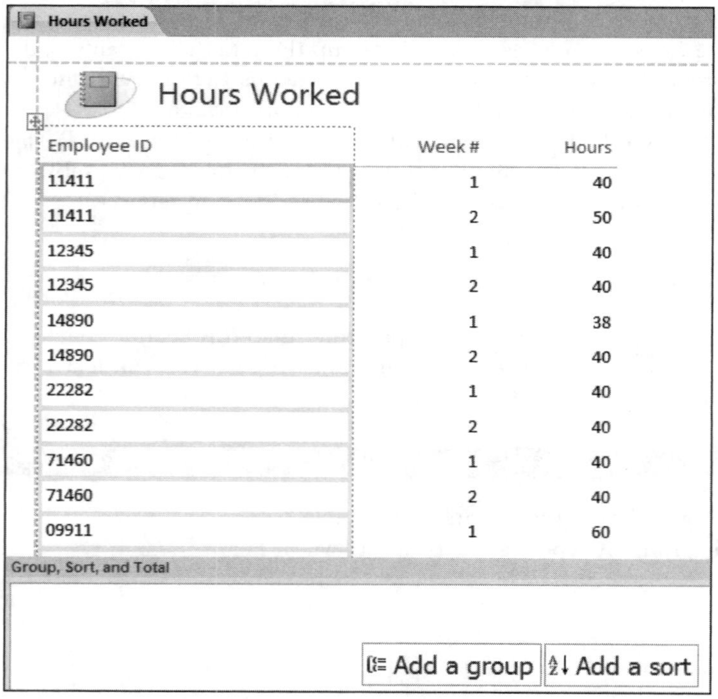

FIGURE B-64   Report with Grouping and Sorting options

Click the Add a group button at the bottom of the report, and then select Employee ID. Your report will be grouped as shown in Figure B-65.

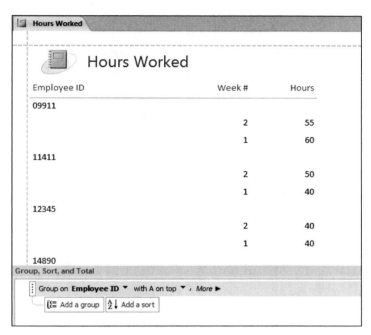

FIGURE B-65 Grouped report

To complete this report, you need to total the hours for each employee by selecting the Hours column heading. Your report will show that the entire column is selected. On the Design tab, click the Totals button in the Grouping and Totals group, and then choose Sum from the menu, as shown in Figure B-66.

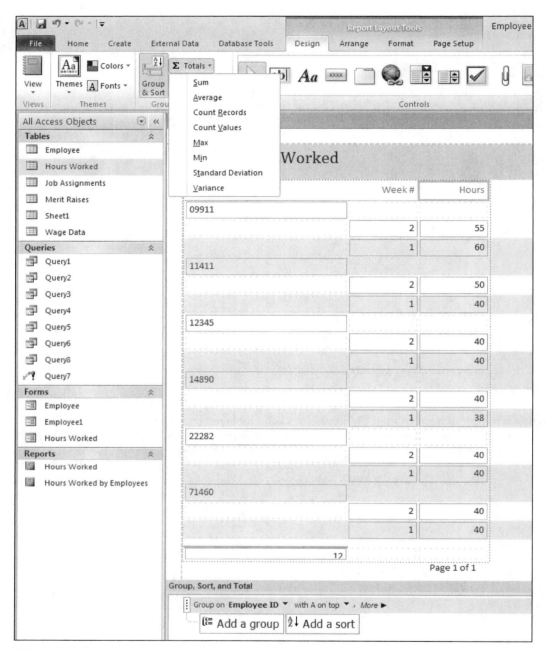

FIGURE B-66 Hours column selected

Your report will look like the one in Figure B-67.

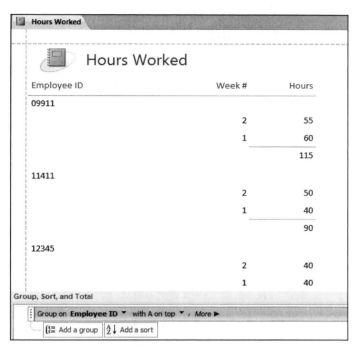

FIGURE B-67 Completed report

Your report is currently in Layout view. To see how the final report looks when printed, click the Design tab and select Report view from the Views group. Your report looks like the one in Figure B-68, although only a portion is shown in the figure.

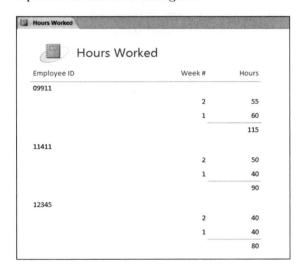

FIGURE B-68 Report in Report view

NOTE

To change the picture or logo in the upper-left corner of the report, simply click the notebook symbol and press the Delete key. You can insert a logo in place of the notebook by clicking the Design tab and then clicking Logo in the Controls group.

Moving Fields in Layout View

If you group on more than one field in a report, the report will have an odd "staircase" look or display repeated data, or it will have both problems. Next, you will learn how to overcome these problems in Layout view.

Suppose you make a query that shows an employee's last name, first name, week number, and hours worked, and then you make a report from that query, grouping on last name only. See Figure B-69.

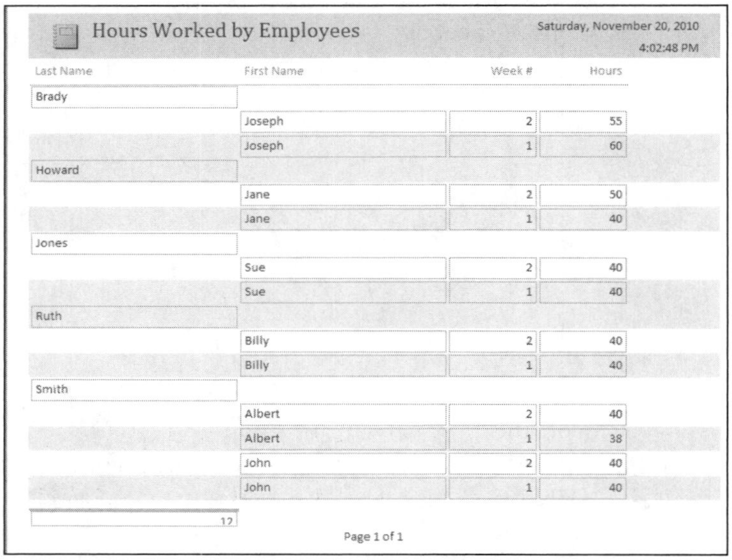
Hours Worked by Employees

Saturday, November 20, 2010 4:02:48 PM

Last Name	First Name	Week #	Hours
Brady			
	Joseph	2	55
	Joseph	1	60
Howard			
	Jane	2	50
	Jane	1	40
Jones			
	Sue	2	40
	Sue	1	40
Ruth			
	Billy	2	40
	Billy	1	40
Smith			
	Albert	2	40
	Albert	1	38
	John	2	40
	John	1	40

12

Page 1 of 1

FIGURE B-69 Query-based report grouped on last name

As you preview the report, notice the repeating data from the First Name field. In the report shown in Figure B-69, notice that the first name repeats for each week worked—hence, the staircase effect. The Week # and Hours fields are shown as subordinate to Last Name, as desired.

Suppose you want the last name and first name to appear on the same line. If so, take the report into Layout view for editing. Click the first record for the First Name (in this case, Joseph), and drag the name up to the same line as the Last Name (in this case, Brady). Your report will now show the First Name on the same line as Last Name, thereby eliminating the staircase look, as shown in Figure B-70.

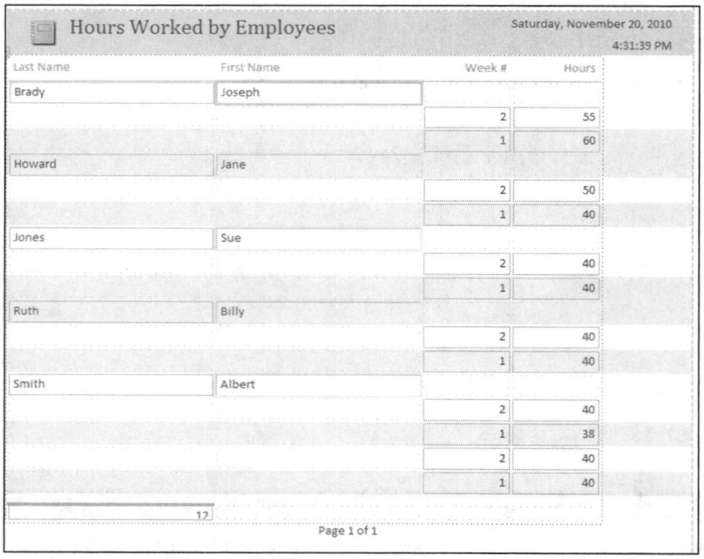
Hours Worked by Employees

Saturday, November 20, 2010 4:31:39 PM

Last Name	First Name	Week #	Hours
Brady	Joseph		
		2	55
		1	60
Howard	Jane		
		2	50
		1	40
Jones	Sue		
		2	40
		1	40
Ruth	Billy		
		2	40
		1	40
Smith	Albert		
		2	40
		1	38
		2	40
		1	40

12

Page 1 of 1

FIGURE B-70 Report with Last Name and First Name on the same line in Layout view

You can now add the sum of Hours for each group. Also, if you want to add more fields to your report, such as Street Address and Zip, you can repeat the preceding procedure.

IMPORTING DATA

Text or spreadsheet data is easy to import into Access. In business, it is often necessary to import data because companies use disparate systems. For example, assume that your healthcare coverage data is on the human resources manager's computer in a Microsoft Excel spreadsheet. Open the Excel application and then create a spreadsheet using the data shown in Figure B-71.

	A	B	C
1	Employee ID	Provider	Level
2	11411	BlueCross	family
3	12345	BlueCross	family
4	14890	Coventry	spouse
5	22282	None	none
6	71460	Coventry	single
7	09911	BlueCross	single

FIGURE B-71   Excel data

Save the file and then close it. Now you can easily import the spreadsheet data into a new table in Access. With your Employee database open, click the External Data tab, then click Excel in the Import & Link group. Browse to find the Excel file you just created, and make sure the first radio button is selected to import the source data into a new table in the current database (see Figure B-72). Click OK.

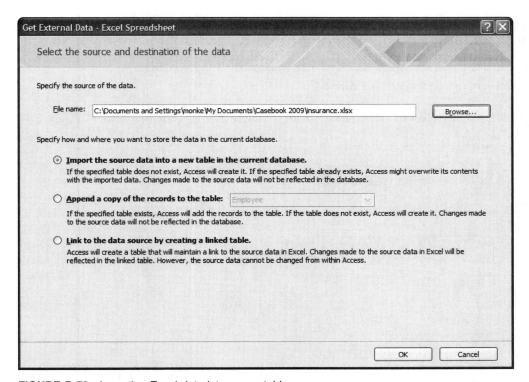

FIGURE B-72   Importing Excel data into a new table

Choose the correct worksheet. Assuming that you have just one worksheet in your Excel file, your next screen should look like the one in Figure B-73.

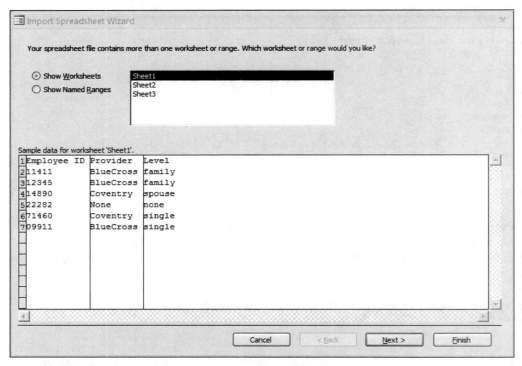

FIGURE B-73   First screen in the Import Spreadsheet Wizard

Choose Next and make sure to select the First Row Contains Column Headings box, as shown in Figure B-74.

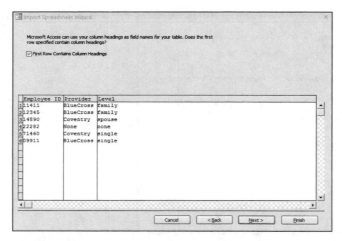

FIGURE B-74   Choosing column headings in the Import Spreadsheet Wizard

Choose Next. Accept the default setting for each field you are importing on the screen. Each field is assigned a text data type, which is correct for this table. Your screen should look like the one in Figure B-75.

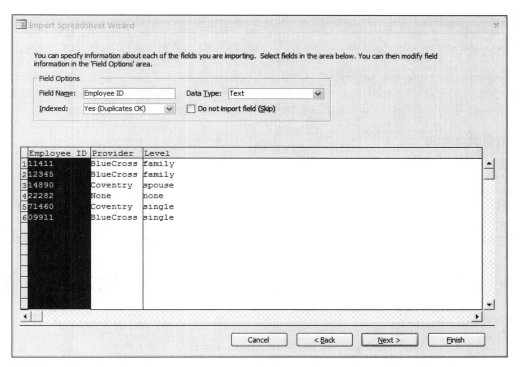

FIGURE B-75 Choosing the data type for each field in the Import Spreadsheet Wizard

Choose Next. In the next screen of the wizard, you will be prompted to create an index—that is, define a primary key. Because you will store your data in a new table, choose your own primary key (Employee ID), as shown in Figure B-76.

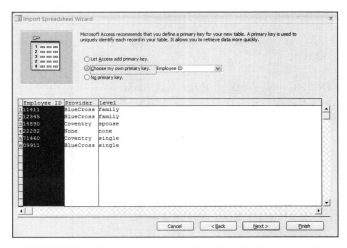

FIGURE B-76 Choosing a primary key field in the Import Spreadsheet Wizard

Continue through the wizard, giving your table an appropriate name. After importing the table, take a look at its design by highlighting the Table option and clicking the Design button. Note that each field is very wide. Adjust the field properties as needed.

MAKING FORMS

Forms simplify the process of adding new records to a table. Creating forms is easy, and can be applied to one or more tables.

When you base a form on one table, you simply select the table, click the Create tab, and then select Form from the Forms group. The form will then contain only the fields from that table. When data is entered into the form, a complete new record is automatically added to the table. Forms with two tables are discussed next.

Making Forms with Subforms

You also can create a form that contains a subform, which can be useful when the form is based on two or more tables. Return to the example Employee database to see how forms and subforms would be useful for viewing all of the hours that each employee worked each week. Suppose you want to show all of the fields from the Employee table; you also want to show the hours each employee worked by including all fields from the Hours Worked table as well.

To create the form and subform, first create a simple one-table form on the Employee table. Follow these steps:

1. Click once to select the Employee table. Click the Create tab, then click Form in the Forms group. After the main form is complete, it should resemble the one in Figure B-77.

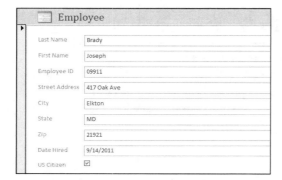

FIGURE B-77   The Employee form

2. To add the subform, take the form into Design view. In the Design tab, make sure that the Use Control Wizards option is selected, scroll to the bottom row of icons in the Controls group, and click the Subform/Subreport icon, as shown in Figure B-78.

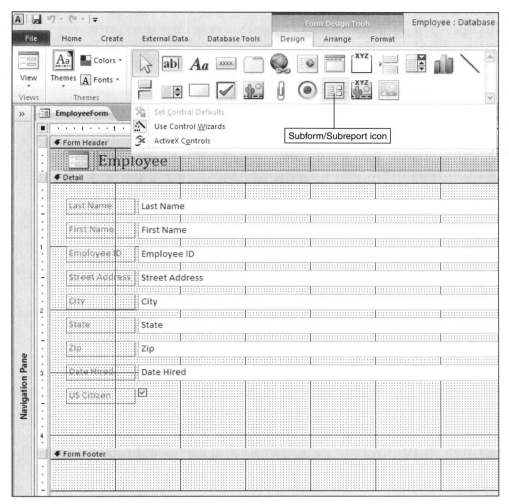

FIGURE B-78   The Subform/Subreport icon

3.   Use your cursor to stretch out the box under your main form. The dialog box shown in Figure B-79 appears.

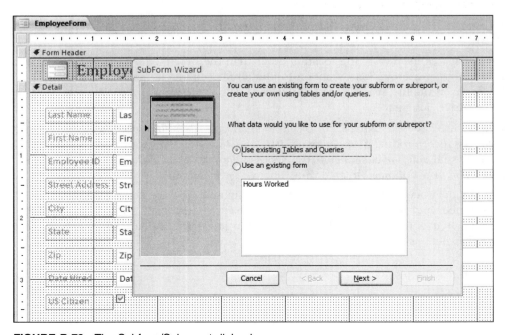

FIGURE B-79   The Subform/Subreport dialog box

4. Select Use existing Tables and Queries, then select Hours Worked from the list. Click Next, select Choose from a List, click Next again, and then click Finished. Select the Form view. Your form and subform should resemble Figure B-80. You may need to stretch out the subform box in Design view if all fields are not visible.

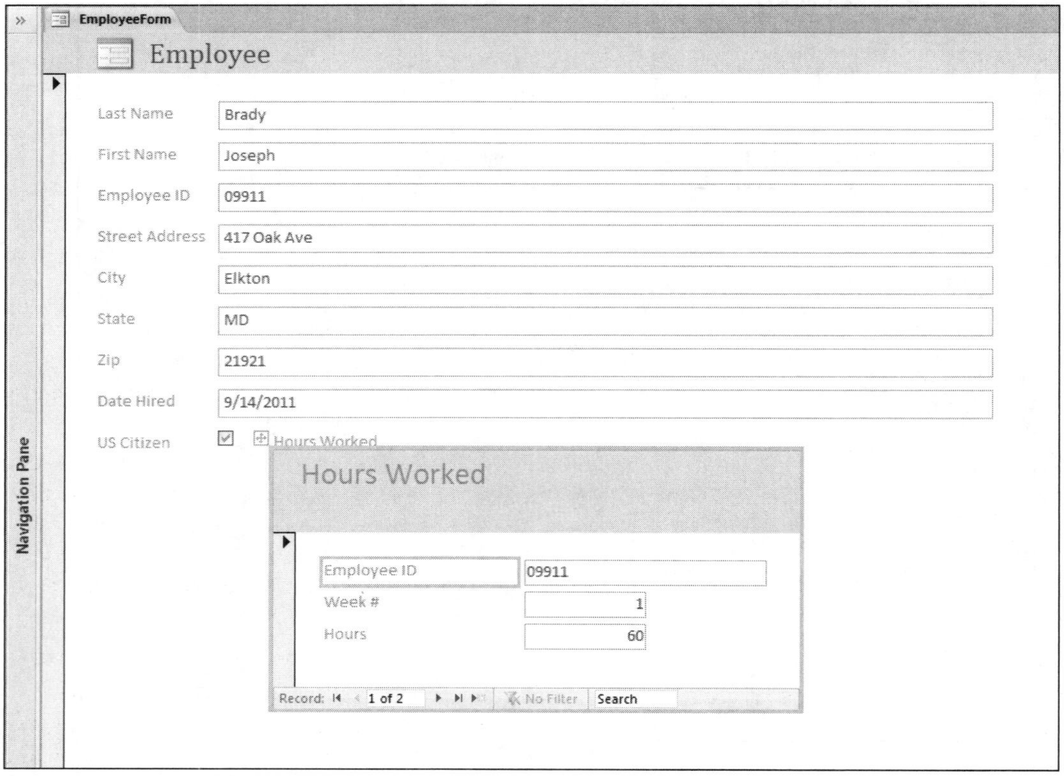

FIGURE B-80   Form with subform

TROUBLESHOOTING COMMON PROBLEMS

Access users sometimes create databases that have problems. Some of these common problems are described below, along with their causes and corrections.

1. *"I saved my database file, but I can't find it on my computer or my external secondary storage medium! Where is it?"*

    You saved your file to a fixed disk or a location other than the Documents folder. Click the Windows Start button, then use the Search option to find all files ending in .accdb (search for *.accdb). If you saved the file, it is on the hard drive (C:\) or a network drive. Your site assistant can tell you the drive designators.

2. *"What is a 'duplicate key field value'? I'm trying to enter records into my Sales table. The first record was for a sale of product X to customer 101, and I was able to enter that one. But when I try to enter a second sale for customer #101, Access tells me I already have a record with that key field value. Am I allowed to enter only one sale per customer?"*

    Your primary key field needs work. You may need a compound primary key—a combination of the customer number and some other field(s). In this case, the customer number, product number, and date of sale might provide a unique combination of values, or you might consider using an invoice number field as a key.

3. *"My query reads 'Enter Parameter Value' when I run it. What is that?"*

    This problem almost always indicates that you have misspelled a field name in an expression in a Criteria field or calculated field. Access is very fussy about spelling; for example, it is case-sensitive. Access is also "space-sensitive," meaning that when you insert a space in a field name when defining a table, you must also include a space in the field name when you reference it in a query expression. Fix the typo in the query expression.

4. *"I'm getting an enormous number of rows in my query output—many times more than I need. Most of the rows are duplicates!"*

    This problem is usually caused by a failure to link all of the tables you brought into the top half of the query generator. The solution is to use the manual click-and-drag method to link the common fields between tables. The spelling of the field names is irrelevant because the link fields need not have the same spelling.

5. *"For the most part, my query output is what I expected, but I am getting one or two duplicate rows or not enough rows."*

    You may have linked too many fields between tables. Usually, only a single link is needed between two tables. It is unnecessary to link each common field in all combinations of tables; it is usually sufficient to link the primary keys. A simplistic explanation for why overlinking causes problems is that it causes Access to "overthink" and repeat itself in its answer.

    On the other hand, you might be using too many tables in the query design. For example, you brought in a table, linked it on a common field with some other table, but then did not use the table. In other words, you brought down none of its fields, and/or you used none of its fields in query expressions. In this case, if you got rid of the table, the query would still work. Click the unneeded table's header at the top of the QBE area, and press the Delete key to see if you can make the few duplicate rows disappear.

6. *"I expected six rows in my query output, but I only got five. What happened to the other one?"*

    Usually, this problem indicates a data entry error in your tables. When you link the proper tables and fields to make the query, remember that the linking operation joins records from the tables *on common values* (*equal* values in the two tables). For example, if a primary key in one table has the value "123," the primary key or the linking field in the other table should be the same to allow linking. Note that the text string "123" is not the same as the text string " 123"—the space in the second string is considered a character too. Access does not see unequal values as an error. Instead, Access moves on to consider the rest of the records in the table for linking. The solution is to examine the values entered into the linked fields in each table and fix any data entry errors.

7. *"I linked fields correctly in a query, but I'm getting the empty set in the output. All I get are the field name headings!"*

    You probably have zero common (equal) values in the linked fields. For example, suppose you are linking on Part Number, which you declared as text. In one field, you have part numbers "001", "002", and "003"; in the other table, you have part numbers "0001", "0002", and "0003". Your tables have no common values, which means that no records are selected for output. You must change the values in one of the tables.

8. *"I'm trying to count the number of today's sales orders. A Totals query is called for. Sales are denoted by an invoice number, and I made that a text field in the table design. However, when I ask the Totals query to 'Sum' the number of invoice numbers, Access tells me I cannot add them up! What is the problem?"*

    Text variables are words! You cannot add words, but you can count them. Use the Count Totals operator (not the Sum operator) to count the number of sales, each being denoted by an invoice number.

9. *"I'm doing time arithmetic in a calculated field expression. I subtracted the Time In from the Time Out and got a decimal number! I expected eight hours, and I got the number .33333. Why?"*

[Time Out] – [Time In] yields the decimal percentage of a 24-hour day. In your case, eight hours is one-third of a day. You must complete the expression by multiplying by 24: ([Time Out] – [Time In]) * 24. Don't forget the parentheses.

10. *"I formatted a calculated field for Currency in the query generator, and the values did show as currency in the query output; however, the report based on the query output does not show the dollar sign in its output. What happened?"*

Go to the report Design view. A box in one of the panels represents the calculated field's value. Click the box and drag to widen it. That should give Access enough room to show the dollar sign as well as the number in the output.

11. *"I told the Report Wizard to fit all of my output to one page. It does print to one page, but some of the data is missing. What happened?"*

Access fits all the output on one page by leaving data out. If you can tolerate having the output on more than one page, deselect the Fit to a Page option in the wizard. One way to tighten output is to enter Design view and remove space from each box that represents output values and labels. Access usually provides more space than needed.

12. *"I grouped on three fields in the Report Wizard, and the wizard prints the output in a staircase fashion. I want the grouping fields to be on one line. How can I do that?"*

Make adjustments in Design view and Layout view. See the "Creating Reports" section of this tutorial for instructions on making these adjustments.

13. *"When I create an Update query, Access tells me that zero rows are updating or more rows are updating than I want. What is wrong?"*

If your Update query is not set up correctly (for example, if the tables are not joined properly), Access will either try not to update anything or it will update all of the records. Check the query, make corrections, and run it again.

14. *"I made a Totals query with a Sum in the Group By row and saved the query. Now when I go back to it, the Sum field reads 'Expression,' and 'Sum' is entered in the field name box. Is that wrong?"*

Access sometimes changes the Sum field when the query is saved. The data remains the same, and you can be assured your query is correct.

15. *"I cannot run my Update query, but I know it is set up correctly. What is wrong?"*

Check the security content of the database by clicking the Security Content button. You may need to enable certain actions.

CASE **1**

PRELIMINARY CASE: VEGGIE BOX DELIVERY

Setting up a Relational Database to Create Tables, Forms, Queries, and Reports

PREVIEW

In this case, you will create a relational database for a local farm that delivers boxes of fresh fruits and vegetables to local customers on a weekly basis. First, you will create three tables and populate them with data. Next, you will create a form and subform for recording new customers and their orders. You will create four queries: a select query, a parameter query, a totals query, and a query used as the basis for a report. Finally, you will create the report from the fourth query.

PREPARATION

- Before attempting this case, you should have some experience using Microsoft Access.
- Complete any part of Access Tutorial B that your instructor assigns, or refer to the tutorial as necessary.

BACKGROUND

More and more people are buying "local" to be more environmentally friendly. In addition, many people feel that they can taste the difference between locally grown produce and produce that is grown far away and shipped to their grocery store. In response, a farm just outside of Atlanta, Georgia is selling boxes of its in-season fruits and vegetables to Atlanta residents. Customers sign up for the so-called "veggie boxes" and pay on a monthly basis. The veggie boxes are delivered to customers' homes every Thursday afternoon. The contents of the box are a surprise; for example, customers do not know if they will receive fresh asparagus, peaches, blueberries, or other fruits and vegetables that were picked that morning.

The veggie box is growing in popularity, and the owners of the farm need your help. The owners have heard that you are proficient in Microsoft Access, and they have asked you to computerize their ordering system. They had hired a summer intern last year, so the database design is already created.

Your first job is to create the tables and populate them with data. Until now, customer orders have been tracked manually. The database design includes three tables, as shown in Figures 1-1, 1-2, and 1-3: Customers, which keeps track of each customer's ID number, name, address, e-mail address, and credit card number for billing purposes; Boxes, which keeps track of the three box types, their description, and monthly price; and Orders, which keeps track of each order number, customer ID, box type, and start date of the service.

The owners have a few requirements for information output in the database beyond simply recording the data. First, they would like to have an easier way to record a new customer's information and the type of box the customer ordered. You can accomplish this task by creating a form and subform.

In addition, the owners would like the database to answer some questions. To enable efficient delivery of the veggie boxes, the drivers want to be able to print the delivery addresses on their routes and order the list by zip code. In addition, the owners want to know how many orders they receive for each type of box so they can plan for next year's crop and subsequent harvest.

The "C" box type contains three vegetables and three fruits. The owners realize that they priced these boxes too cheaply, so they want to notify all customers who ordered the "C" boxes that the price will be increased by 20 percent next year. The owners want a listing of these customers so they can prepare a mass e-mail that explains the price increase. Finally, the owners want to be able to produce a list of bills for a particular month. Ideally, this information should be stored in a nicely formatted report.

ASSIGNMENT 1: CREATING TABLES

Use Microsoft Access to create the tables with the fields shown in Figures 1-1 through 1-3; these tables were discussed in the Background section. Populate the database tables as shown. Add your name to the Customers table with an appropriate customer ID; complete the entry by adding your address, phone number, e-mail address, and a fictional credit card number.

Customer ID	Last Name	First Name	Address	City	State	Zip	Telephone	Cell Phone	Email Address	Credit Card	Click to Add
B-17	Bianco	Anna	9 Pleasant Way	Atlanta	GA	30600	404-887-4673	404-876-3376	aberry@hotmail.com	443376562837	
F-59	Franklin	Gina	1012 Peachtree St	Atlanta	GA	30600	404-887-2342	404-765-1263	gf59@gmail.com	443398764532	
L-29	LaGoia	Maria	5490 West 5th	Atlanta	GA	30600	404-234-8876	569-001-0989	mrl@hotmail.com	443352635423	
M-62	McMillan	Annabelle	59 W. Central Ave	Atlanta	GA	30600	404-998-3928	404-887-3829	belle@comcast.net	443355463212	
P-91	Prince	Jill	89 Orchard	Atlanta	GA	30600	404-887-9238	404-342-9087	pao@comcast.net	443367256543	
Q-13	Queller	Sally	54 Oak Ave	Atlanta	GA	30600	404-987-3427	569-984-3894	quinn45@gmail.com	443398765439	
S-63	Sandy	Patricia	1700 E. Lincoln Ave	Atlanta	GA	30600	404-765-3342	404-121-4736	patti1@gmail.com	443398762534	
Z-30	Zepher	Joan	58 W. Central Ave	Atlanta	GA	30600	404-675-0091	404-776-4536	zern@comcast.net	443357643254	

FIGURE 1-1   The Customers table

Box Type	Description	Monthly Price	Click to Add
A	Two vegetables, Two fruits	$28.50	
B	Three vegetables, Two fruits	$35.40	
C	Three vegetables, Three fruits	$38.75	

FIGURE 1-2   The Boxes table

Order Number	Customer ID	Box Type	Start Date	Click to Add
101	B-17	A	4/1/2012	
102	Z-30	B	6/1/2012	
103	F-59	A	5/1/2012	
104	S-63	C	4/1/2012	
105	L-29	C	5/1/2012	
106	Q-13	B	7/1/2012	
107	M-62	B	8/1/2012	
108	P-91	B	4/1/2012	

FIGURE 1-3   The Orders table

ASSIGNMENT 2: CREATING A FORM, QUERIES, AND A REPORT

Assignment 2A: Creating a Form

Create a form for easy recording of new customers and their orders. The main form should be based on the Customers table, and the subform should be inserted with the fields from the Orders table. Save the form as Customers. View one record and, if required by your instructor, print the record. Your output should resemble that in Figure 1-4.

Customers	
Customers	
Customer ID	B-17
Last Name	Bianco
First Name	Anna
Address	9 Pleasant Way
City	Atlanta
State	GA
Zip	30600
Telephone	404-887-4673
Cell Phone	404-876-3376
Email Address	aberry@hotmail.com
Credit Card	443376562837

Order Number	Customer	Box Type	Start Date
101	B-17	A	4/1/2012
*	B-17		

Record: 2 of 2   No Filter   Search

FIGURE 1-4   The Customers form with subform

Assignment 2B: Creating a Parameter Query

Create a parameter query that prompts for a zip code and subsequently lists columns for the Last Name, First Name, Address, Zip Code, and Box Type of all customers in the respective zip code. Save your query as Delivery by Zip Code. Your output should resemble Figure 1-5 when you enter the zip code 30600 at the prompt.

Delivery by Zip Code

Last Name	First Name	Address	Zip	Box Type
Bianco	Anna	9 Pleasant Way	30600	A
Zepher	Joan	58 W. Central Ave	30600	B
Franklin	Gina	1012 Peachtree St	30600	A
Sandy	Patricia	1700 E. Lincoln Ave	30600	C
LaGoia	Maria	5490 West 5th	30600	C
Queller	Sally	54 Oak Ave	30600	B
McMillan	Annabelle	59 W. Central Ave	30600	B
Prince	Jill	89 Orchard	30600	B
*				

FIGURE 1-5   Delivery by Zip Code query

Run the query. Print the results if required.

Assignment 2C: Creating a Totals Query

Create a query that adds up the number of boxes on order for the month of July. Your output should show the box type, description, monthly price, and number of orders for July. (Note that the Number of Orders heading is a column heading change from the default setting provided by the query generator.) Save your query as July Order Summary. Your output should resemble that shown in Figure 1-6. Print the output if desired.

July Order Summary

Box Type	Description	Monthly Price	Number of Orders
A	Two vegetables, Two fruits	$28.50	2
B	Three vegetables, Two fruits	$35.40	3
C	Three vegetables, Three fruits	$38.75	2

FIGURE 1-6 July Order Summary query

Assignment 2D: Creating a Query with a Calculated Field

Create a query that calculates the new price of a "C" box, which is 20 percent higher than the current price. Include columns that list the Last Name, First Name, and Email Address of all "C" box recipients, along with the new price. Save the query as Price Increase Notification. Your output should resemble that shown in Figure 1-7. Print the output if desired.

Price Increase Notification

	Last Name	First Name	Email Address	New Price
	Sandy	Patricia	patti1@gmail.com	$46.50
	LaGoia	Maria	mrl@hotmail.com	$46.50
*				

FIGURE 1-7 Price Increase Notification query

Assignment 2E: Generating a Report

Generate a report based on a query. The query should display the last name, first name, and address of all customers who receive veggie boxes in April, along with the monthly price and start date. Save the query as April Bills. From that query, create a report with a grand total of all monthly payments at the bottom. Make sure that all fields and data are visible, and title the report April Bills. Your report output should resemble that in Figure 1-8.

April Bills

April Bills

Monday, November 08, 2010
2:40:42 PM

Last Name	First Name	Address	Monthly Price	Start Date
Bianco	Anna	9 Pleasant Way	$28.50	4/1/2012
Sandy	Patricia	1700 E. Lincoln Ave	$38.75	4/1/2012
Prince	Jill	89 Orchard	$35.40	4/1/2012
			$102.65	

Page 1 of 1

FIGURE 1-8 April Bills report

If you are working with a portable storage disk or USB key, make sure that you remove it *after* closing the database file.

DELIVERABLES

Assemble the following deliverables for your instructor, either electronically or in printed form:

1. Three tables
2. Form and subform: Customers
3. Query 1: Delivery by Zip Code
4. Query 2: July Order Summary
5. Query 3: Price Increase Notification
6. Query 4: April Bills
7. Report: April Bills
8. Any other required tutorial printouts or electronic media

Staple all pages together. Put your name and class number at the top of the page. If required, make sure that your electronic media are labeled.

INTERNET JETS RESERVATION SYSTEM DATABASE

Designing a Relational Database to Create Tables, Forms, Queries, and Reports

PREVIEW

In this case, you will design a relational database for a private jet leasing service. After your database design is completed and correct, you will create database tables and populate them with data. Then you will produce one form with a subform that allows you to reserve jets; you will also produce five queries and one report. The queries will address the following questions: Which members live in a specified state? What are the reservations for mid-March? What is the most popular jet reserved? Who is the most popular pilot in terms of customer requests? Your report, based on a query, will display the reservations requested by each member.

PREPARATION

- Before attempting this case, you should have some experience in database design and in using Microsoft Access.
- Complete any part of Database Design Tutorial A that your instructor assigns.
- Complete any part of Access Tutorial B that your instructor assigns, or refer to the tutorial as necessary.
- Refer to Tutorial F as necessary.

BACKGROUND

Internet Jets is a company that provides private airplane service to customers who have joined the organization and pay a yearly fee. The growing company is interested in having you design and create a database for its reservation system. Billing and payments will be executed in a different system that is already up and running.

Before you begin, you need to know a few details about the business. Internet Jets operates by recruiting members to actually own a small part of the company's fleet of airplanes. For example, members pay a yearly fee to own 1/16th of an airplane, and are then able to fly for approximately 50 hours per year. The benefits of this approach include predictable costs and a guaranteed jet for use whenever the need arises.

Your system should include information about the members of this cooperative company. Each member will be assigned a unique identification number to avoid erroneous bookings. Other recorded information will include the member's name, address, cell phone number, and e-mail address, because all bookings are reserved via the Internet.

Internet Jets owns three different categories of airplanes: models with light cabins, mid-sized cabins, and large cabins. Figure 2-1 shows the capacity and range of each type of plane.

Type	Capacity	Range
Light cabin	7 people	1600 miles
Mid-sized cabin	8 people	3000 miles
Large cabin	10 people	4300 miles

FIGURE 2-1   Types of jets

When a booking comes into Internet Jets, it is displayed like a form is displayed in Microsoft Access. The booking information shows the type of jet requested, the customer, the origination and destination of the trip, and the date and time requested. The booking agent at Internet Jets assigns an airplane to each request. Ignore this step in the process—your design and implementation involve only the reservation requests, not the actual flights assigned and flown.

The jets are flown by professional pilots, and each flight has a captain and co-captain. Some customers like certain pilots and want those pilots to fly their plane. In response to that request, the booking form now includes a space for customers to request a particular pilot. Your form should include a drop-down menu that lists all the pilots so customers can simply click on their requested pilot. In addition, to avoid any mistakes in data entry, the form should include a drop-down menu that lists the various types of available airplanes.

Besides the reservation form, management would like you to create a number of queries to show some important information. First, management wants to know which states that members of the flying cooperative live in so the company can better market and advertise the service. You suggest a query that will prompt for the state as an input; management can then use this query to find members in any state.

Management often wants to know the bookings at different times of the year, and has asked you to create a query that lists the bookings for a particular period of weeks or days. In addition, management is curious about the popularity of certain aspects of the jet service. They would like to know which airplane size is the most popular, and which pilots are the most frequently requested.

Finally, management would like to see a report that lists all customers and the reservations they have made.

ASSIGNMENT 1: CREATING THE DATABASE DESIGN

In this assignment, you will design your database tables using a word-processing program. Pay close attention to the tables' logic and structure. Do not start developing your Access code in Assignment 2 before getting feedback from your instructor on Assignment 1. Keep in mind that you will need to examine the requirements in Assignment 2 to design your fields and tables properly. It is good programming practice to look at the required outputs before beginning your design. When designing the database, observe the following guidelines:

- First, determine the tables you will need by listing the name of each table and the fields it should contain. Avoid data redundancy. Do not create a field if it can be created by a "calculated field" in a query.
- You will need transaction tables. Think about what business events occur with each member's actions. Avoid duplicating data.
- Document your tables using the table feature of your word processor. Your tables should resemble the format shown in Figure 2-2.
- You must mark the appropriate key field(s) by entering an asterisk (*) next to the field name.
- Print the database design if your instructor requires it.

Table Name	
Field Name	Data Type (text, numeric, currency, etc.)
...	...
...	...

FIGURE 2-2   Table design

NOTE

Have your design approved before beginning Assignment 2; otherwise, you may need to redo Assignment 2.

ASSIGNMENT 2: CREATING THE DATABASE, QUERIES, AND REPORT

In this assignment, you will first create database tables in Access and populate them with data. Next, you will create a form, queries, and a report.

Assignment 2A: Creating Tables in Access

In this part of the assignment, you will create your tables in Access. Use the following guidelines:

- Create at least 10 members of the jet leasing service, including yourself as a member, and record the three types of airplanes. You can make up fictional names, addresses, and phone numbers for members, but use the types of airplanes listed in Figure 2-1.
- Create records for four pilots, using your friends' names as fictional pilots.
- Each member should reserve an airplane flight, and at least four members should reserve more than one.
- Appropriately limit the size of the text fields; for example, a telephone number does not need the default length of 255 characters.
- Print all tables if your instructor requires it.

Assignment 2B: Creating Forms, Queries, and a Report

You will generate one form with a subform, five queries, and one report, as outlined in the Background section of this case.

Form

Create a form and subform based on your Members table and Reservations table (or whatever you named these tables). Save the form as Members and Reservations. Your form should resemble that in Figure 2-3.

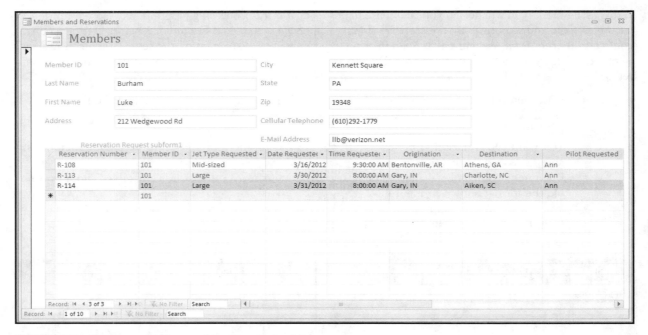

FIGURE 2-3   Members and Reservations form and subform

Query 1

Create a query called Members in What State? In Figure 2-4, the queried state is Pennsylvania (PA). This query should prompt the user to enter a state name. The output of the query should list the last name, first name, full address, and cell phone number of members in the designated state. Your output should resemble that shown in Figure 2-4, although your data will be different.

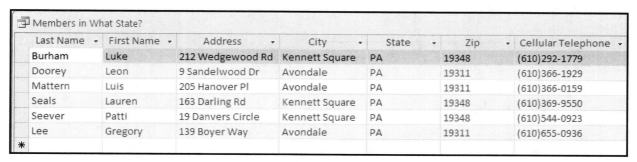

Last Name	First Name	Address	City	State	Zip	Cellular Telephone
Burham	Luke	212 Wedgewood Rd	Kennett Square	PA	19348	(610)292-1779
Doorey	Leon	9 Sandelwood Dr	Avondale	PA	19311	(610)366-1929
Mattern	Luis	205 Hanover Pl	Avondale	PA	19311	(610)366-0159
Seals	Lauren	163 Darling Rd	Kennett Square	PA	19348	(610)369-9550
Seever	Patti	19 Danvers Circle	Kennett Square	PA	19348	(610)544-0923
Lee	Gregory	139 Boyer Way	Avondale	PA	19311	(610)655-0936

FIGURE 2-4   Members in What State? query

Query 2

Create a query called Mid-March Reservations (or use another month if your data is for a different time period). List the date requested, the type of airplane, and the origination of all flights for a particular 10-day period. Your output should look like that in Figure 2-5, although your data will be different.

Date Requested	Jet Type Requested	Origination
3/12/2012	Light	Salt Lake City, UT
3/13/2012	Light	Salt Lake City, UT
3/14/2012	Light	Salt Lake City, UT
3/14/2012	Mid-sized	Ft. Collins, CO
3/15/2012	Mid-sized	Gary, IN
3/16/2012	Mid-sized	Bentonville, AR
3/18/2012	Large	Atlantic City, NJ
3/19/2012	Large	Lexington, VA

FIGURE 2-5   Mid-March Reservations query

Query 3

Create a query called Number of Jets Reserved. In this query, you need to determine the number of times each type of jet has been reserved. List the jet type and the number of times it has been reserved, showing the most popular to the least popular. Note the column heading change from the default setting provided by the query generator. Your output should resemble the format shown in Figure 2-6, but the data will be different.

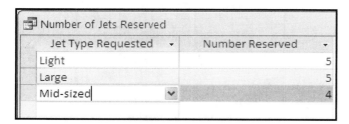

Jet Type Requested	Number Reserved
Light	5
Large	5
Mid-sized	4

FIGURE 2-6   Number of Jets Reserved query

Query 4

Create a query called Most Popular Pilot. In this query, you need to determine the number of times each pilot has been requested. List the name of the requested pilot and the number of times the pilot has been requested, from the most popular to the least popular. It is acceptable to list only the pilot's first name because there are only four pilots, and you can assume that none have the same first name. Note the column heading change from the default setting provided by the query generator. Your output should resemble the format shown in Figure 2-7, but the data will be different.

Most Popular Pilot	
Pilot Requested	Number of Times Requested
Ann	6
Bob	4
Gary	3
Luke	1

FIGURE 2-7   Most Popular Pilot query

Report

Create a report named Reservations by Customer. The basis of this report is a query that displays the first and last name of the member, the type of jet requested, and the date it was requested. Save the query as For Report. Using all the data in the output of the query, create a grouped report. Group the report on Last Name, and bring the First Name to the same grouping line as Last Name. Give the report a title of Reservations by Customer, just like the report's filename. Make sure that all data and column headings are visible. Depending on your data, your output should resemble that shown in Figure 2-8.

Reservations by Customer

Reservations by Customer   Monday, November 08, 2010
                                                          2:56:40 PM

Last Name	First Name	Jet Type Requested	Date Requested
Burham	Luke		
		Large	3/31/2012
		Large	3/30/2012
		Mid-sized	3/16/2012
Doorey	Leon		
		Mid-sized	3/1/2012
Lee	Gregory		
		Mid-sized	3/15/2012
Mattern	Luis		
		Large	3/18/2012
Meartz	Maria		
		Light	3/24/2012
Poplawski	Meredith		
		Mid-sized	3/14/2012
Seals	Lauren		
		Large	3/5/2012
Seever	Patti		
		Light	3/20/2012
Ward	Harry		
		Light	3/14/2012

FIGURE 2-8   Reservations by Customer report

ASSIGNMENT 3: MAKING A PRESENTATION

Create a presentation that explains the database to the management of Internet Jets. Include the design of your database tables and instructions for using the database. Discuss future improvements to the database, such as porting the form to the Internet. Your presentation should take fewer than 10 minutes, including a brief question-and-answer period.

DELIVERABLES

Assemble the following deliverables for your instructor, either electronically or in printed form:

1. Word-processed design of tables
2. Tables created in Access
3. Form and subform: Members and Reservations
4. Query 1: Members in What State?
5. Query 2: Mid-March Reservations
6. Query 3: Number of Jets Reserved
7. Query 4: Most Popular Pilot
8. Query 5: For Report
9. Report: Reservations by Customer
10. Presentation materials
11. Any other required tutorial printouts or electronic media

Staple all pages together. Put your name and class number at the top of the page. Make sure that your electronic media are labeled, if required.

CASE **3**

THE INTRAMURAL SPORTS DATABASE

Designing a Relational Database to Create Tables, Forms, Queries, and Reports

PREVIEW

In this case, you will design a relational database for a paper-based system that tracks students who participate in intramural sports at the local university. After your design is complete and correct, you will populate the tables with data and create a form, queries, and a report. The form will be used for adding new members to the teams. The queries will address the following questions: Which students have not had a medical exam? Which students are members of a particular team? How long do the games last? How many students are on each team? Which students have quit a team, meaning they can be deleted from the active database? Finally, you will create a report that lists the amount of time each referee has worked in the intramural games.

PREPARATION

- Before attempting this case, you should have some experience in database design and in using Microsoft Access.
- Complete any part of Database Design Tutorial A that your instructor assigns.
- Complete any part of Access Tutorial B that your instructor assigns, or refer to the tutorial as necessary.
- Refer to Tutorial F as necessary.

BACKGROUND

You are a student at a local university and have landed a part-time job with the intramural sports department because of your expertise in database design and implementation. The intramural sports department has been keeping track of its members, teams, and events on paper, and now it is drowning in paperwork. You have been hired to design a database and later implement it in Microsoft Access.

On the first day at your job, you interview John, the director of the intramural program. John tells you what types of information he has been keeping on paper and how he uses the information.

John keeps an enormous three-ring binder that contains a tab for each intramural sport. For example, there are tabs for squash, disk golf, and swimming. Each tabbed section has a cover sheet that lists the name of the sport, the names of the intramural teams, and the faculty adviser. Each team is named for a color, and each color is unique. If John needs to get in touch with a team's faculty adviser, he opens his address book, which is ordered by last name, and finds the adviser's telephone number. In all the years that John has been directing the program, he has never had two faculty advisers with the same last name.

Each cover sheet is followed by a preprinted sheet that is filled out for each student on an intramural team for the particular sport. The sheet contains personal information about the students, including local and home addresses and health insurance numbers. John is careful to record each student's identification number because some students have the same name. The bottom of the sheet includes a space to note whether the student has had a medical exam this year.

Keep in mind that some students have joined more than one intramural team in different sports.

When John needs to schedule games or tournaments, he uses a large wall calendar to record the name of the sport, the names of the teams, and the time of day and venue where the event will occur. When the event is over, John records the ending time and puts a gold star next to the winning team's name on the same calendar.

Once a good database is designed and implemented, John can envision plenty of information he can glean from the system. For example, John often has wanted a list of all students who have not had a medical exam. You suggest that a query will answer that question. In addition, the team captains constantly ask for an updated list of their team members and their phone numbers. This list often must be printed immediately before a game because students join and quit intramural teams frequently. You suggest that a parameter query could be effective in this situation; the captain could enter his team's color and get a current listing of team members.

John is curious to know how long each game lasts, because he has received some comments about long games. He would also like to know how many students are on each team. Again, you realize that the answers to both questions could be captured by queries. John would also like to be able to add students to teams and remove them from teams. You tell John that a form and subform could be used for additions, and that a delete query would be appropriate to remove a student from a team.

Finally, you suggest creating a report that displays the amount of time each referee has worked in intramural games. John can use this information to thank the refs by giving them awards, gifts, or pay.

ASSIGNMENT 1: CREATING THE DATABASE DESIGN

In this assignment, you will design your database tables using a word-processing program. Pay close attention to the tables' logic and structure. Do not start developing your Access code in Assignment 2 before getting feedback from your instructor on Assignment 1. Keep in mind that you will need to examine the requirements in Assignment 2 to design your fields and tables properly. It is good programming practice to look at the required outputs before beginning your design. When designing the database, observe the following guidelines:

- First, determine the tables you will need by listing the name of each table and the fields it should contain. Avoid data redundancy. Do not create a field if it can be created by a "calculated field" in a query.
- You will need a transaction table. Although no money is changing hands in this system, events are occurring, and they are considered transactions. Avoid duplicating data.
- Consider using a logical field to note whether each student has had a medical exam.
- Document your tables using the table feature of your word processor. Your tables should resemble the format shown in Figure 3-1.
- You must mark the appropriate key field(s) by entering an asterisk (*) next to the field name. Keep in mind that some tables might need a compound primary key to uniquely identify a record within a table.
- Print the database design.

Table Name	
Field Name	Data Type (text, numeric, currency, etc.)
...	...
...	...

FIGURE 3-1 Table design

NOTE

Have your design approved before beginning Assignment 2; otherwise, you may need to redo Assignment 2.

ASSIGNMENT 2: CREATING THE DATABASE, QUERIES, AND REPORT

In this assignment, you will first create database tables in Access and populate them with data. Next, you will create a form with a subform, several queries, and a report.

Assignment 2A: Creating Tables in Access

In this part of the assignment, you will create your tables in Access. Use the following guidelines:

- Create at least six sports teams with at least three different sports.
- Create records for at least 10 students and have them join multiple teams.
- Make up at least 10 games, and vary the beginning time and ending time of each.
- Appropriately limit the size of the text fields; for example, a telephone number does not need the default length of 255 characters.
- Print all tables.

Assignment 2B: Creating Forms, Queries, and a Report

You will generate one form, six queries, and one report, as outlined in the Background section of this case.

Form

Create a form and subform based on your Students table and Team Members table (or whatever you named these tables). Save the form as Students and Team Members. Your form should resemble that in Figure 3-2.

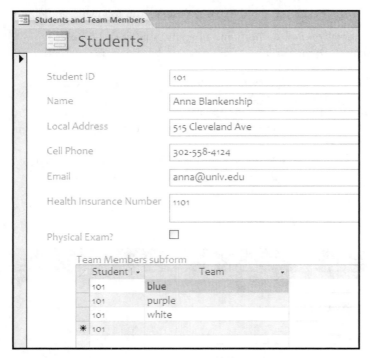

FIGURE 3-2   Students and Team Members form and subform

Query 1

Create a query called Students without Physical Exam. Include headings that list the Name, Local Address, and Cell Phone of students who have not had medical exams. Your output should resemble that in Figure 3-3, although your data will differ.

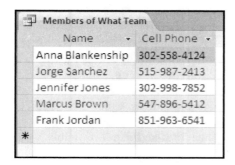

FIGURE 3-3   Students without Physical Exam query

Query 2

Create a query called Members of What Team. This query should prompt the user to enter a team color, and then list the names and cell phone numbers of the team members. Your output will differ, but the layout should resemble that in Figure 3-4.

FIGURE 3-4   Members of What Team query

Query 3

Create a query called Game Times that calculates the length of each game and shows the date and game time. Your data will differ, but the output should resemble that in Figure 3-5.

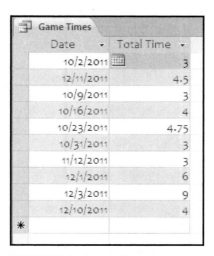

FIGURE 3-5   Game Times query

Query 4

Create a query called Number of Students on Teams. The query should list team names and the number of players on each team. Sort the output to this query. Your output should resemble that in Figure 3-6, although the data will differ. Note the column heading change from the default setting provided by the query generator.

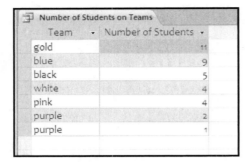

FIGURE 3-6   Number of Students on Teams query

Query 5

Create a delete query that prompts the user to enter the ID number of a student whose intramural record should be deleted. The query should also prompt the user to enter the student's associated team color. Save the query as Delete Members. Test the query by entering a new record and deleting it.

Report

Create a report called Referee Times. Begin the report by creating a query that displays columns for the Referee Last Name, Referee First Name, Date, Game Time, and Winning Team. (The Game Time is the total time of the game, which is a calculation.) Save the query as For Report, bring the query into a report, and group on Referee Last Name. The report should include subtotals of game times for each referee. Adjust the output so that the Referee First Name is on the same line as the Referee Last Name. Your output should resemble that in Figure 3-7. Make sure that all headings and data are visible.

Referee Last Name	Referee First Name	Date	Game Time	Winning Team
Jacobsen	Paula			
		12/11/2011	4.5	pink
		12/10/2011	4	pink
		10/31/2011	3	purple
		10/23/2011	4.75	gold
			16.25	
Smith	Joe			
		12/3/2011	9	black
		12/1/2011	6	black
		11/12/2011	3	purple
		10/16/2011	4	blue
		10/9/2011	3	blue
		10/2/2011	3	blue
			28	

Referee Times — Monday, November 15, 2010 — 2:31:07 PM

FIGURE 3-7   Referee Times report

ASSIGNMENT 3: MAKING A PRESENTATION

Create a presentation for John and his colleagues. Describe your database design and the design of your form, queries, and report. Show samples of the outputs. Discuss future expansion of the system. Your presentation should take fewer than 10 minutes, including a brief question-and-answer period.

DELIVERABLES

Assemble the following deliverables for your instructor, either electronically or in printed form:

1. Word-processed design of tables
2. Tables created in Access
3. Form: Students and Team Members
4. Query 1: Students without Physical Exam
5. Query 2: Members of What Team
6. Query 3: Game Times
7. Query 4: Number of Students on Teams
8. Query 5: Delete Members
9. Query 6: For Report
10. Report: Referee Times
11. Presentation materials
12. Any other required tutorial printouts or electronic media

Staple all pages together. Put your name and class number at the top of the page. Make sure that your electronic media are labeled, if required.

Create a PDF report combined if possible!

Home
View
Print Preview
Publish

T-SHIRTS ETC. ORDER DATABASE

Designing a Relational Database to Create Tables, Forms, Queries, and Reports

PREVIEW

In this case, you will design a relational database for a business that prints custom t-shirts and assorted clothing and sells them to customers who attend the local university. After your database design is completed and correct, you will create database tables and populate them with data. Then you will produce one form with a subform, six queries, and two reports. The queries will address the following questions: What items sell for less than $20? What are the best-selling products? What are the most popular colors in the best-selling products? Who are the best customers? Another query will help the owner update prices on items. One report will list customer orders, including specific discounts based on a query. Another report will display the output from a query more professionally.

PREPARATION

- Before attempting this case, you should have some experience in database design and in using Microsoft Access.
- Complete any part of Database Design Tutorial A that your instructor assigns.
- Complete any part of Access Tutorial B that your instructor assigns, or refer to the tutorial as necessary.
- Refer to Tutorial F as necessary.

BACKGROUND

Susan is a local artist who has years of experience in the textile industry. Her children attend the local university, and they sometimes ask her to design and print custom t-shirts for their university organizations, such as clubs or sports teams. Friends of Susan's children have always complimented them on the custom shirts, and have requested some of their own. Susan has decided to go into business printing custom apparel and selling it to organizations at the local university. Her forecasts indicate that her first-year sales will be quite high, so she will be very busy generating all the products for the orders. As a consequence, she wants to hire a part-time worker to handle the orders that come into the shop. One of Susan's children is a friend of yours, and knows you have mastered database design and implementation. Your friend asks you to apply for the job. Before you begin, you interview Susan about how the company works.

YOU: Tell me about how you take orders.

SUSAN: When a potential customer requests a custom job, I first take down all the customer information, such as the person's name, street address, city, state, and zip code. I also request their phone number in case I have questions about the order.

YOU: Sorry to interrupt, but can some customers have the same name?

SUSAN: Yes. As I was saying, then I show them the list of items I have for customization in our showroom, which is my dining room. There I show them t-shirts, both long and short sleeved; fleece jackets, either pullover or zipped; hoodies, caps, and fleece hats. Each has a different price, which is marked on the item, and each can be ordered in a variety of colors and sizes, except headwear. When the customer is ready to order, I fill out a form with the customer information repeated on the top of the order. I would like not to have to rewrite all this information because I sometimes make copying mistakes. Then I put the date at the top and usually include an order number that I have created. After that heading, I list the item the customer wants, including size, color, and quantity.

YOU: What other information would you like to get out of the database?

SUSAN: I would like to be able to take orders quickly, because certain times are very busy for me. In addition, my market is college students, who are often on a limited budget. I would like to be able to show them which items are cheaper than $20. I would also like a way to change prices easily, because the cost of textiles is rising fast.

YOU: I can create a form for you to enter the orders quickly, and queries will help you with the other two tasks. I can also help you analyze your sales with specific queries that might help you sell more in the future. Can you think of some marketing analysis that would be helpful?

SUSAN: I would love to see a listing of the best-selling products, the best-selling colors, and a list of who my best customers have been. That way, I can send them coupons for future work and hope to increase sales.

YOU: How about creating the bills to send out to customers?

SUSAN: I would like to have a list of all my customers with their orders individually noted, and a summary of what they owe me. I could pass the list on to my collection agency, which handles all my unpaid bills. In addition, I often give a 10% discount to customers in the next city, Glendale, to create more repeat business.

ASSIGNMENT 1: CREATING THE DATABASE DESIGN

In this assignment, you will design your database tables using a word-processing program. Pay close attention to the tables' logic and structure. Do not start developing your Access code in Assignment 2 before getting feedback from your instructor on Assignment 1. Keep in mind that you will need to examine the requirements in Assignment 2 to design your fields and tables properly. It is good programming practice to look at the required outputs before beginning your design. When designing the database, observe the following guidelines:

- First, determine the tables you will need by listing the name of each table and the fields it should contain. Avoid data redundancy. Do not create a field if it can be created by a "calculated field" in a query.
- You will need transaction tables. Think about what business events occur with each customer's actions. Avoid duplicating data.
- Document your tables using the table feature of your word processor. Your tables should resemble the format shown in Figure 4-1.
- You must mark the appropriate key field(s) by entering an asterisk (*) next to the field name. Keep in mind that some tables might need a compound primary key to uniquely identify a record within a table.
- Print the database design.

Table Name	
Field Name	Data Type (text, numeric, currency, etc.)
...	...
...	...

FIGURE 4-1   Table design

> **NOTE**
>
> Have your design approved before beginning Assignment 2; otherwise, you may need to redo Assignment 2.

ASSIGNMENT 2: CREATING THE DATABASE, QUERIES, AND REPORT

In this assignment, you will first create database tables in Access and populate them with data. Next, you will create a form, six queries, and two reports.

Assignment 2A: Creating Tables in Access

In this part of the assignment, you will create your tables in Access. Use the following guidelines:

- Enter at least seven records for the products: two types of t-shirts (short and long sleeved), two types of fleece jackets (pullover and zipped), hoodies, baseball caps, and fleece hats. Assume that all apparel except headwear comes in small, medium, and large sizes.
- Enter records for at least nine customers. Use your own name, address, telephone number, and e-mail address to create an additional customer record. Assume that the business comes from your university town and one other town nearby.
- Each customer should have at least one order; a few customers should place two orders. Each order should contain multiple items.
- Appropriately limit the size of the text fields; for example, a telephone number does not need the default length of 255 characters.
- Print all tables if your instructor requires it.

Assignment 2B: Creating Forms, Queries, and Reports

You will generate one form with a subform, six queries, and two reports, as outlined in the Background section of this case.

Form

Create a form and subform based on your Orders table and Order Line Item table (or whatever you named these tables). Save the form as Orders. Your form should resemble that in Figure 4-2.

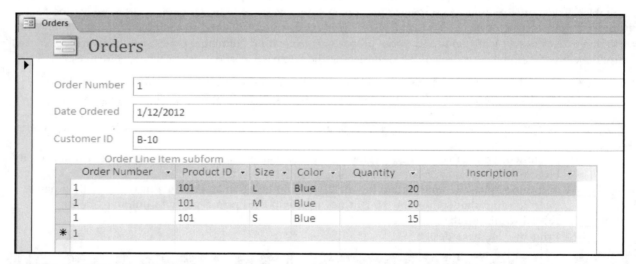

FIGURE 4-2   Orders form and subform

Query 1

Create a select query called Products Less Than $20 that displays a list of all products that cost less than $20. Your output should resemble that shown in Figure 4-3, although your data will be different.

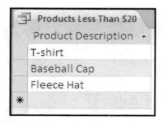

FIGURE 4-3   Products Less Than $20 query

Query 2

Create a parameter/update query that prompts for the incremental amount of a price increase and then prompts for the product description. Save the query as Updated Prices. Test the query by adding $0.25 to the cost of a product. View the changed price in your datasheet view of the updated table.

Query 3

Create a query called Favorite Colors. List the colors of the products ordered and determine how many have been ordered of each; report the amounts in a column labeled Number Ordered. Sort the query output. Note the column heading change from the default setting provided by the query generator. Your output should resemble the format shown in Figure 4-4, but the data will be different.

Favorite Colors	
Color	Number Ordered
Pink	144
Yellow	131
Blue	131
Purple	125
Brown	120
Grey	90
Red	50
Black	25
White	15
Green	15

FIGURE 4-4 Favorite Colors query

Query 4

Create a query called Best Selling Product. List the product descriptions and determine how many orders have been received for each product. Report the amounts in a column labeled Number Ordered. Sort the output. Note the column heading change from the default setting provided by the query generator. Your output should resemble the format shown in Figure 4-5, but the data will be different.

Best Selling Product	
Product Description	Number Ordered
T-shirt, long-sleeve	311
T-shirt	175
Fleece Hat	125
Baseball Cap	100
Fleece Jacket	56
Fleece Pullover	55
Hoodie	24

FIGURE 4-5 Best Selling Product query

Query 5

Create a query called Best Customers that lists the customers' names, phone numbers, and the total amounts of their orders. You will have to calculate the order amounts and sort the output. Note the column heading change from the default setting provided by the query generator. Your output should resemble the format shown in Figure 4-6, but the data will be different.

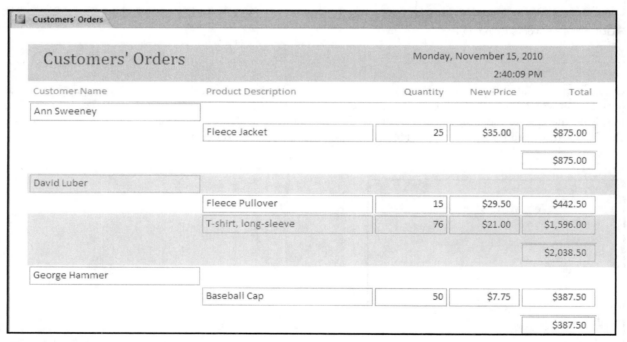

FIGURE 4-6   Best Customers query

Report 1

Create a report named Customers' Orders. The report's output should show headings for Customer Name, Product Description, Quantity, New Price, and Total. All of this data originates in a query, in which all customers who live in a specified city receive a 10 percent discount. All other customers pay the full price. The discounted price or full price is noted in the New Price column. Then you calculate the amount of money owed for each product, which is the New Price multiplied by the Quantity. Save the query as For Report, bring the query data into a report, and group the report on Customer Name. Make sure that all column headings and data are visible and that all money amounts are formatted properly into currency. Depending on your data, your output should resemble that shown in Figure 4-7.

FIGURE 4-7   Customers' Orders report

Report 2

To impress Susan, bring the Favorite Colors query output into a report. Save the report as Favorite Colors. Make sure that all column headings and data are visible. Depending on your data, your report should resemble that in Figure 4-8.

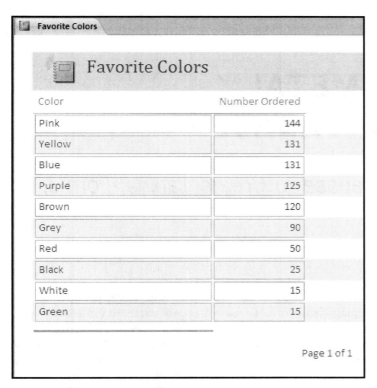

FIGURE 4-8 Favorite Colors report

ASSIGNMENT 3: MAKING A PRESENTATION

Create a presentation that explains the database to Susan. Include the design of your database tables and instructions for using the database. Discuss future improvements to the database, such as placing product information and the order form on the Web. Your presentation should take fewer than 10 minutes, including a brief question-and-answer period.

DELIVERABLES

Assemble the following deliverables for your instructor, either electronically or in printed form:

1. Word-processed design of tables
2. Tables created in Access
3. Form and subform: Orders
4. Query 1: Products Less Than $20
5. Query 2: Updated Prices
6. Query 3: Favorite Colors
7. Query 4: Best Selling Product
8. Query 5: Best Customers
9. Query 6: For Report
10. Report 1: Customers' Orders
11. Report 2: Favorite Colors
12. Presentation materials
13. Any other required tutorial printouts or electronic media

Staple all pages together. Put your name and class number at the top of the page. Make sure that your electronic media are labeled, if required.

THE PRECIOUS METAL DEPOSITORY DATABASE

Designing a Relational Database to Create Tables, Forms, Queries, and Reports

PREVIEW

In this case, you will design a tracking system relational database for a company that stores precious metals for investors. After your database design is completed and correct, you will create database tables and populate them with data. Then you will produce two forms, nine queries, and two reports. The forms will record all deliveries and withdrawals from the depository. The queries will list the guards who worked at the depository on a specific day, determine the most popular deliveries and withdrawals, and calculate the hours worked by the guards for a specified month. Queries will also determine the city wage tax for a specific month. In addition, queries will list deliveries and withdrawals for customers by precious metal type and shape (bullion bars or coins). You will use a query to delete erroneous records. Finally, you will create two reports to list guards' hours for specific months.

PREPARATION

- Before attempting this case, you should have some experience in database design and in using Microsoft Access.
- Complete any part of Database Design Tutorial A that your instructor assigns.
- Complete any part of Access Tutorial B that your instructor assigns, or refer to the tutorial as necessary.
- Refer to Tutorial F as necessary.

BACKGROUND

The Precious Metal Depository, known as PMD, is located in a nondescript building in suburban Los Angeles. PMD holds precious metal bars called bullion and precious metal coins for investors. In the past, investors would simply hold a certificate or "note" that represented their ownership of an expensive metal. Now, given the credit crunch and shaky financial markets, investors often want to own the metals themselves. PMD serves these customers by holding their precious metals in its highly secure building. Only a handful of companies specialize in this type of business.

The price of precious metals has risen significantly as investors pursue stable returns. Investing in precious metals such as gold, silver, platinum, and palladium is usually considered safer in times of financial turmoil. Precious metals usually come either in bullion or coins.

PMD believes in the "3 Ss"—safety, security, and sensitivity—as its top priorities. The company's armored trucks are always unmarked, and the drivers have undergone extensive background checks; many were trained in the military. Vaults at the depository are heavily secured, and two guards are always present when the bullion and coins are handled. Customers are assured that their data is secure and that no sensitive information is released.

PMD has proprietary software to track its customer records and the precious metals. PMD is fairly confident that the software accurately tracks the extensive movement of bullion and coins in and out of the facility. However, PMD would like an additional system that tracks this movement and the guards on duty while the precious metals are being moved. You have been hired as a summer intern to design a small database system that records the hours each guard works at PMD and records the movement of precious metals in and out of the building.

The owner of PMD, Jerome Argent, begins by explaining aspects of the business that are important to your database design. Guards work around the clock at PMD, patrolling the main entrance to the building. Once a person enters the highly secure building, he or she must use biometric identification to enter any vault area. After a person enters the vestibule of the vault, the entry door must be closed before the door to the vault is opened. At least two guards are on duty at all times. All guards have undergone background checks. Information about the guards, such as their names, addresses, and phone numbers, is kept on record. Two guards might have the same name. – Emp. ID

Guards clock in using a time card system before each shift and clock out afterward. Guards are paid on an hourly basis. They watch the movement of precious metals in and out of the building, and they record information about each delivery and withdrawal. This information includes the customer number (names are kept secret), the amount deposited or withdrawn, and the time. PMD prides itself on accurately recording the movement of goods.

Next, Jerome explains the different products stored in the vaults. Each precious metal comes in two formats: bullion, which is shaped like a brick or bar, and coins. The precious metals bought and sold are gold, silver, platinum, and palladium.

Jerome has a list of requirements for the database system. First, when guards record the movement of goods in and out of the building, they should have a computerized method of entering the information rather than writing it down. You suggest forms as a vehicle for this input.

Other information can be gleaned from queries. For example, Jerome would like to see which guards are working at specific dates and times, in case any suspicious activity or problems are reported. He is also curious to know which precious metals are the most popular. For example, which metal is in greatest supply at the depository? For each metal, which is more popular: bullion or coins? Which metal is most frequently delivered, and which is most frequently withdrawn? Jerome would like the database to be able to break down deliveries and withdrawals by metal type and by shape (bullion or coins). You suggest a parameter query for each of these questions. The queries can prompt the user to specify a type of metal and specify whether it is held in bullion or coins.

Jerome would also like you to calculate the hours worked by each guard over a specified period of months, one month at a time. In addition, he wants you to calculate the guards' gross pay for a month and compute the city tax for guards who live in Los Angeles. You confidently state that queries can handle these requests, including calculations. You also suggest that some queries might be displayed through the report generator to provide a more polished output and to group the data by guard.

Finally, Jerome tells you that some customers cancel withdrawals at the last minute, depending on how the market is trading that day. As the withdrawal is being executed and the guard is recording the particulars, it might be reversed. Jerome would like a fast way to delete a withdrawal recorded by a guard.

ASSIGNMENT 1: CREATING THE DATABASE DESIGN

In this assignment, you will design your database tables using a word-processing program. Pay close attention to the tables' logic and structure. Do not start developing your Access code in Assignment 2 before getting feedback from your instructor on Assignment 1. Keep in mind that you will need to examine the requirements in Assignment 2 to design your fields and tables properly. It is good programming practice to look at the required outputs before beginning your design. When designing the database, observe the following guidelines:

- First, determine the tables you will need by listing the name of each table and the fields it should contain. Avoid data redundancy. Do not create a field if it can be created by a "calculated field" in a query.
- You will need a number of transaction tables. Think about the business events that will occur. Avoid duplicating data.

- Keep in mind that all customers are known only by ID numbers for security and privacy reasons.
- Document your tables using the table feature of your word processor. Your tables should resemble the format shown in Figure 5-1.
- You must mark the appropriate key field(s) by entering an asterisk (*) next to the field name. Keep in mind that some tables might need a compound primary key to uniquely identify a record within a table.
- Print the database design.

Table Name	
Field Name	Data Type (text, numeric, currency, etc.)
...	...
...	...

FIGURE 5-1   Table design

NOTE

Have your design approved before beginning Assignment 2; otherwise, you may need to redo Assignment 2.

ASSIGNMENT 2: CREATING THE DATABASE, QUERIES, AND REPORTS

In this assignment, you will first create database tables in Access and populate them with data. You will create a table that records delivery data and another that records withdrawal data. Next, you will create two forms, nine queries, and two reports.

Assignment 2A: Creating Tables in Access

In this part of the assignment, you will create your tables in Access. Use the following guidelines:

- Enter 10 records for guards in the tables. Use the names and addresses of your friends and relatives to create records for the guards.
- Assume that PMD holds four precious metals: gold, silver, platinum, and palladium. Each type comes in either bullion or coins.
- Create a large number of transactions for metals being delivered and withdrawn. Consider using Excel to generate data for some of the tables. You might choose to use the =randbetween function in Excel to generate random numbers within a range you specify. You then can import the Excel-generated data into a new Access table. (For details, see the section on importing data in Tutorial B.)
- Create a large number of hours worked for the guards. Again, consider generating plenty of data in Excel.
- Appropriately limit the size of the text fields; for example, a zip code does not need the default length of 255 characters.
- Print all tables if your instructor requires it.

Assignment 2B: Creating Forms, Queries, and Reports

You will create two forms, nine queries, and two reports, as outlined in the Background section of this case.

Form 1

Create a form based on your table that records delivery data. Use all the fields in your table to create the form. Name the form Deliveries. Your data will vary, but the output should resemble that in Figure 5-2.

Deliveries

Deliveries

Delivery ID	123456
Customer ID	2150
Product ID	2
Quantity	9
Date	11/15/2011
Time	10:45:00 AM

FIGURE 5-2   Deliveries form

Form 2

Create a form based on your table that records withdrawal data. Use all the fields in your table to create the form. Name the form Withdrawals. Your data will vary, but the output should resemble that in Figure 5-3.

Withdrawals

Withdrawals

Withdrawal ID	654321
Customer ID	2071
Product ID	6
Quantity	6
Date	11/12/2011
Time	3:15pm

FIGURE 5-3   Withdrawals form

Query 1

Create a query called Guard on New Year's Day. Display headings for the Employee Last Name, Employee First Name, and Date to summarize which guards worked on New Year's Day. (If necessary, you can specify a different day depending on your data.) Your data will differ, but your output should resemble that in Figure 5-4.

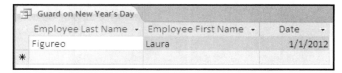

Employee Last Name	Employee First Name	Date
Figureo	Laura	1/1/2012

FIGURE 5-4   Guard on New Year's Day query

Query 2

Create a query called Popular Deliveries that displays headings for the Metal Type, Shape, and Number Delivered. Be sure to sort your output. Your data will differ, but your output should resemble that in Figure 5-5. Note the column heading change from the default setting provided by the query generator.

Metal Type ▾	Shape ▾	Number Delivered ▾
Silver	Coin	95
Palladium	Bullion	87
Platinum	Coin	85
Platinum	Bullion	63
Gold	Bullion	63
Palladium	Coin	55
Gold	Coin	53
Silver	Bullion	52

Popular Deliveries

FIGURE 5-5   Popular Deliveries query

Query 3

Create a query called Popular Withdrawals that displays headings for the Metal Type, Shape, and Number Withdrawn. Be sure to sort your output. Your data will differ, but your output should resemble that in Figure 5-6. Note the column heading change from the default setting provided by the query generator.

Metal Type ▾	Shape ▾	Number Withdrawn ▾
Platinum	Coin	97
Silver	Coin	82
Silver	Bullion	82
Palladium	Coin	81
Gold	Coin	76
Palladium	Bullion	68
Platinum	Bullion	48
Gold	Bullion	37

Popular Withdrawals

FIGURE 5-6   Popular Withdrawals query

Query 4

Create a query called Deliveries by Type and Shape. Display headings for the Customer ID, Number Delivered, Metal Type, and Shape. This query should prompt for two inputs: Metal Type and Shape. Your data will differ, but your output should resemble that in Figure 5-7 if you entered Gold and Bullion at the prompts.

Deliveries by Type and Shape

Customer ID	Number Delivered	Metal Type	Shape
2887	10	Gold	Bullion
2981	8	Gold	Bullion
2627	8	Gold	Bullion
2537	8	Gold	Bullion
2471	8	Gold	Bullion
2858	4	Gold	Bullion
2645	4	Gold	Bullion
2796	3	Gold	Bullion
2177	3	Gold	Bullion
2613	2	Gold	Bullion
2176	2	Gold	Bullion
2263	1	Gold	Bullion
2149	1	Gold	Bullion
2080	1	Gold	Bullion

FIGURE 5-7 Deliveries by Type and Shape query

Query 5

Create a query called Withdrawals by Type and Shape. Display headings for the Customer ID, Number Withdrawn, Metal Type, and Shape. This query should prompt for two inputs: Metal Type and Shape. Your data will differ, but your output should resemble that in Figure 5-8 if you entered Gold and Bullion at the prompts.

Withdrawals by Type and Shape

Customer ID	Number Withdrawn	Metal Type	Shape
2339	10	Gold	Bullion
2151	10	Gold	Bullion
2948	6	Gold	Bullion
2483	5	Gold	Bullion
2836	4	Gold	Bullion
2546	2	Gold	Bullion

FIGURE 5-8 Withdrawals by Type and Shape query

Query 6

Create a query called November Hours that summarizes the shifts worked by guards during November. Display headings for the Employee Last Name, Employee First Name, Date, Number of Hours (a calculated field), and Employee ID. Your data will differ, but your output should resemble that in Figure 5-9. Note that only a portion of the output is showing in the figure.

Between 11/1/2011 And 11/30/2011

November Hours				
Employee Last Name ▾	Employee First Name ▾	Date ▾	Number of Hours ▾	Employee ID ▾
Katz	Karen	11/12/2011	10	3
Wright	Danny	11/13/2011	10	7
Buchanan	Patricia	11/14/2011	10	6
Brown	Mark	11/15/2011	10	5
Watson	Wayne	11/16/2011	10	9
Katz	Karen	11/17/2011	10	3
Wright	Danny	11/18/2011	10	7
Baker	Marie	11/19/2011	10	2
McGregor	Francis	11/20/2011	10	4
Baker	Marie	11/12/2011	6	2
Brown	Mark	11/13/2011	6	5
Figureo	Laura	11/14/2011	6	10
Wright	Danny	11/15/2011	6	7
Abraham	Isaac	11/16/2011	6	8

FIGURE 5-9   November Hours query

Query 7

Create a query called November LA Tax, using the November Hours query and an additional table as input. Assume that all guards earn $20 per hour, and that guards who live in Los Angeles have an additional one percent tax on their income. Include both of these assumptions in the query, using an IF statement to calculate the tax for Los Angeles residents. List headings for the Employee Last Name, Employee First Name, Pay, and City Tax. Your data will differ, but your output should resemble that in Figure 5-10. Note that only a portion of the output is showing in the figure.

Numbers of Hrs • 20

November LA Tax			
Employee Last Name ▾	Employee First Name ▾	Pay ▾	City Tax ▾
Katz	Karen	$200.00	$2.00
Wright	Danny	$200.00	$2.00
Buchanan	Patricia	$200.00	$2.00
Brown	Mark	$200.00	$0.00
Watson	Wayne	$200.00	$0.00
Katz	Karen	$200.00	$2.00
Wright	Danny	$200.00	$2.00
Baker	Marie	$200.00	$2.00
McGregor	Francis	$200.00	$0.00
Baker	Marie	$120.00	$1.20
Brown	Mark	$120.00	$0.00
Figureo	Laura	$120.00	$0.00

FIGURE 5-10   November LA Tax query

Query 8

Create a query called December Hours that summarizes the shifts worked by guards during December. Display headings for the Employee Last Name, Employee First Name, Date, and Number of Hours (a calculated field). Your data will differ, but your output should resemble that in Figure 5-11. Note that only a portion of the output is showing in the figure.

Employee Last Name	Employee First Name	Date	Number of Hours
Buchanan	Patricia	12/1/2011	10
Buchanan	Patricia	12/2/2011	10
Figureo	Laura	12/3/2011	10
Baker	Marie	12/4/2011	10
Baker	Marie	12/5/2011	10
Buchanan	Patricia	12/6/2011	10
Abraham	Isaac	12/7/2011	10
Katz	Karen	12/8/2011	10
Ortiz	Jorge	12/9/2011	10
Figureo	Laura	12/10/2011	10
Figureo	Laura	12/11/2011	10
Buchanan	Patricia	12/12/2011	10

FIGURE 5-11   December Hours query

Query 9

Create a delete query called Delete Withdrawal. This query deletes a record from the Withdrawals table after prompting the user to enter a customer ID and date. Test your query by running it for a record and then looking at the Withdrawals table to make sure the record was deleted.

Report 1

Create a report called November Hours that is based on the November Hours query, but only displays headings for the Employee Last Name, Employee First Name, Date, and Number of Hours. Group the report on Employee Last Name. Adjust the output so that the Employee First Name is on the same line as the Employee Last Name. Sum the Number of Hours for each guard. Your data will differ, but your output should resemble that in Figure 5-12. Include November Hours as the title of the report. All data and headings should be visible, although only a portion of the report is showing in Figure 5-12.

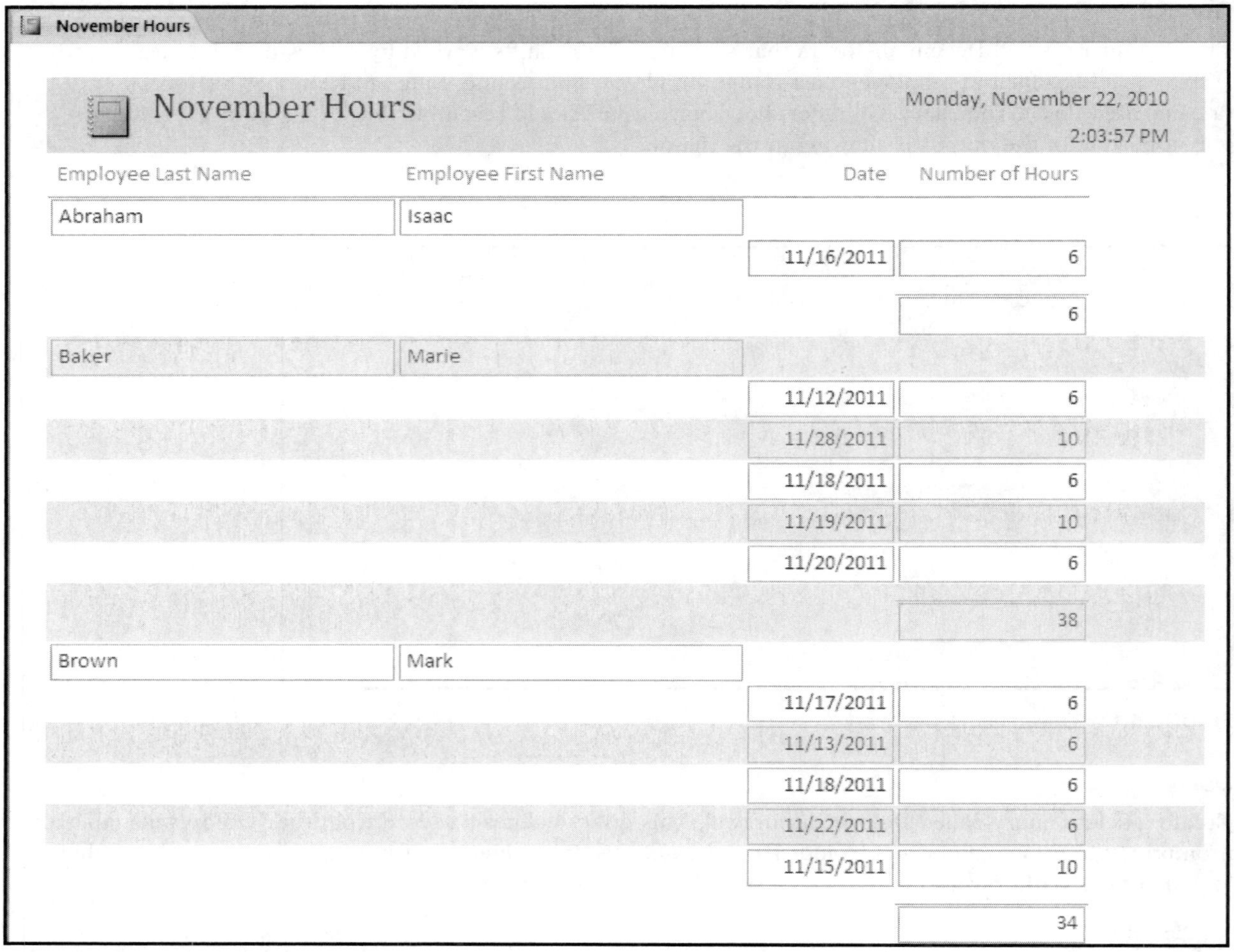

FIGURE 5-12 November Hours report

Report 2

Create a report called December Hours that is based on the December Hours query. Group the report on Employee Last Name. Adjust the output so that the Employee First Name is on the same line as the Employee Last Name. Sum the Number of Hours for each guard. Your data will differ, but your output should resemble that in Figure 5-13. Include December Hours as the title of the report. All data and headings should be visible, although only a portion of the report is showing in Figure 5-13.

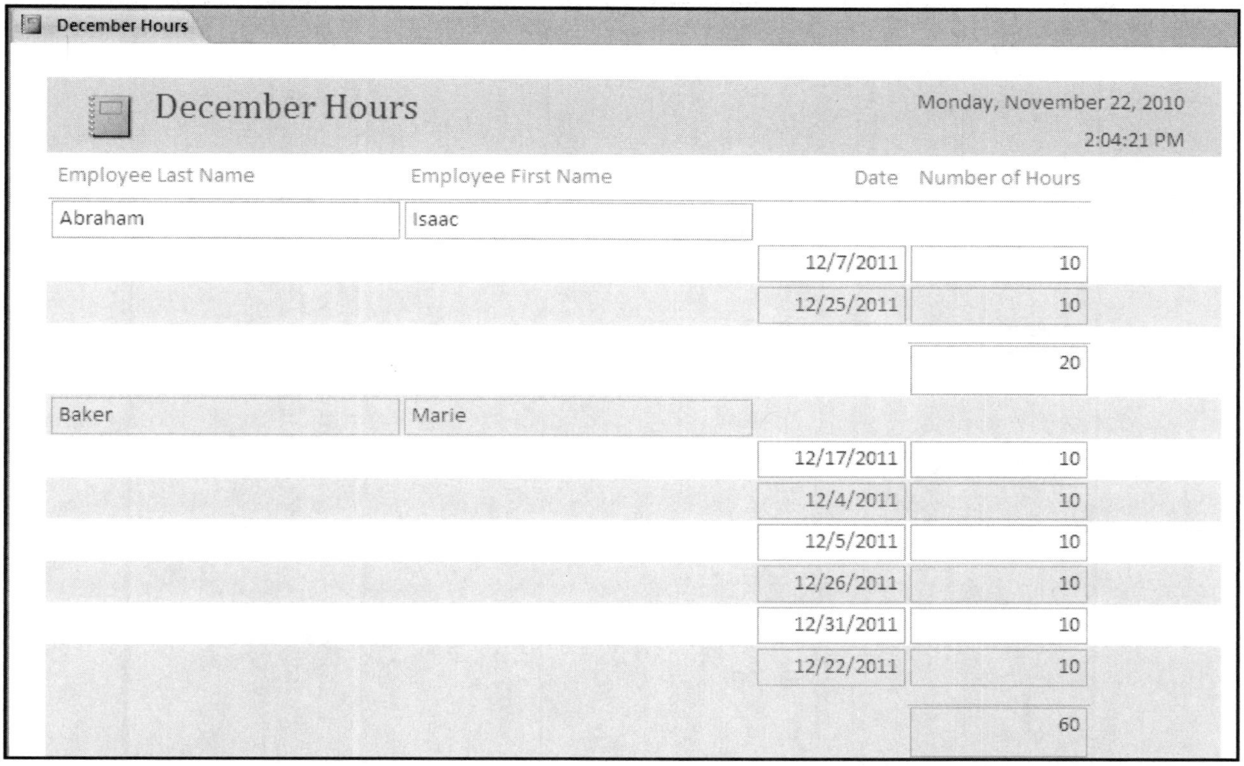

FIGURE 5-13   December Hours report

ASSIGNMENT 3: MAKING A PRESENTATION

Create a presentation for the Precious Metal Depository. Pay particular attention to potential database users who are not familiar with Microsoft Access. Discuss computer and data security in your presentation and discuss how the project could be expanded. Your presentation should take fewer than 15 minutes, including a brief question-and-answer period.

DELIVERABLES

Assemble the following deliverables for your instructor, either electronically or in printed form:

1. Word-processed design of tables
2. Tables created in Access
3. Form 1: Deliveries
4. Form 2: Withdrawals
5. Query 1: Guard on New Year's Day
6. Query 2: Popular Deliveries
7. Query 3: Popular Withdrawals
8. Query 4: Deliveries by Type and Shape
9. Query 5: Withdrawals by Type and Shape
10. Query 6: November Hours
11. Query 7: November LA Tax
12. Query 8: December Hours
13. Query 9: Delete Withdrawal
14. Report 1: November Hours
15. Report 2: December Hours

Staple all pages together. Put your name and class number at the top of the page. Make sure that your electronic media are labeled, if required.

PART 2

DECISION SUPPORT CASES USING EXCEL SCENARIO MANAGER

TUTORIAL **C**

BUILDING A DECISION SUPPORT SYSTEM IN EXCEL

Decision Support Systems (DSS) are computer programs used to help managers solve complex business problems. DSS programs are commonly found in large, integrated packages called enterprise resource planning software that provide information services to an organization. Software packages such as SAP™, Microsoft Dynamics™, and PeopleSoft™ offer sophisticated DSS capabilities. However, many business problems can be modeled for solutions using less complex tools such as Visual Basic, Access, and Microsoft Excel.

A DSS program is actually a model representing a quantitative business problem. The problem can range from finding a desired product mix to sales forecasts to risk analysis, but almost all of the problems examine *financial outcomes*. The model itself contains the data and the algorithms (mathematical processes) needed to solve the problem.

In a DSS program, either the user manually inputs data or the program accesses data from a file in the system. The program runs the data through its algorithms and displays output formatted as information; the manager uses this data to decide what action to take to solve the problem. Some sophisticated DSS programs display multiple possible solutions and recommend one based on predefined parameters.

Managers often find the Excel spreadsheet program particularly useful for their DSS needs. Excel contains hundreds of built-in arithmetic, statistical, logical, and financial functions. It can import data in numerous formats from large database programs, and it can be set up to display well-organized, visually appealing tables and graphs from the output.

This tutorial is organized into four sections:

1. **Spreadsheet and DSS Basics**—This section lets you "get your feet wet" by creating a DSS program in Excel. The program is a cash flow model for a small business looking to expand. You will get an introduction to spreadsheet design, building a DSS, and using financial functions.
2. **Scenario Manager**—Here you will learn how to use the Excel Scenario Manager. A DSS typically gives you one set of answers based on one set of inputs—the real value of the tool lies in its ability to play "what if" and take a comparative look at all the solutions based on all combinations of the inputs. Rather than inputting and running the DSS several times manually, you can use Scenario Manager to run and display the outputs from all possible combinations of the inputs. The output is summarized on a separate worksheet in the Excel workbook.
3. **Practice Using Scenario Manager**—Next, you will be given a new problem to model as a DSS, using Scenario Manager to display your solutions.
4. **Review of Excel Basics**—This section reviews additional information that will help you complete the spreadsheet cases that follow this tutorial. You will learn some basic operations, logical functions, and cash flow calculations.

SPREADSHEET AND DSS BASICS

You are the owner of a thrift shop that resells clothing and housewares in a university town. Many of your customers are college students. Your business is unusual in that sales actually increase during an economic recession. Your cost of obtaining used items basically follows the consumer price index. It is the end of 2011, and business has been very good due to the continuing recession. You are thinking of expanding your business to an adjacent storefront that is for sale, but you will have to apply for a business loan to finance the purchase. The bank will require a projection of your profit and cash flows for the next two years before it will loan you the money to expand, so you have to determine your net income (profit) and cash flows for 2012 and 2013. You decide that your forecast should be based on four factors: your 2011 sales dollars, your cost of goods sold per sales dollar, your estimates of the underlying economy, and the business loan payment amount and interest rate.

Because you will present this model to your prospective lenders, you decide to use an Income and Cash Flow Statements framework. You will input values for two possible states of the economy for 2012 and 2013: R for a continuing recession and B for a "boom" (recovery). Your sales in the recession were growing at 20% per year. If the recession continues and you expand the business, you expect sales to continue growing at 30% per year. However, if the economy recovers, some of your customers will switch to buying "new," so you expect sales growth for your thrift shop to be 15% above the previous year (only 5% growth plus 10% for the business expansion). If you do not expand, your recession or boom growth percentages will only be 20% and 5%, respectively. To determine the cost of goods sold for purchasing your merchandise, which is currently 70% of your sales, you will input values for two possible consumer price outlooks: H for high inflation (1.06 multiplied by the average cost of goods sold) and L for low inflation (1.02 multiplied by the cost of goods sold).

You currently own half the storefront and will need to borrow $100,000 to buy and renovate the other half. The bank has indicated that, depending on your forecast, it may be willing to loan you the money for your expansion at 5% interest during the current recession with a 10-year repayment compounded annually ("R"). However, if the prime rate drops at the start of 2012 because of an economic turnaround ("B"), the bank can drop your interest rate to 4% with the same repayment terms.

As an entrepreneur, an item of immediate interest is your cash flow position with the additional burden of a loan payment. After all, one of your main objectives is to make a profit (Net Income After Taxes). You can use the DSS model to determine if it is more profitable *not* to expand the business.

Organization of the DSS Model

A well-organized spreadsheet will make the design of your DSS model easier. Your spreadsheet should have the following sections:

- Constants
- Inputs
- Summary of Key Results
- Calculations (with separate calculations for Expansion vs. No Expansion)
- Income and Cash Flow Statements (with separate statements for Expansion vs. No Expansion)

Figures C-1 and C-2 illustrate the spreadsheet setup for the DSS model you want to build.

	A	B	C	D
1	**Tutorial Exercise--Collegetown Thrift Shop**			
2				
3	**Constants**	**2011**	**2012**	**2013**
4	Tax Rate	NA	33%	35%
5	Loan Amount for Store Expansion	NA	$100,000	NA
6				
7	**Inputs**	**2011**	**2012**	**2013**
8	Economic Outlook (R=Recession, B=Boom)	NA		NA
9	Inflation Outlook (H=High, L=Low)	NA		NA
10				
11	**Summary of Key Results**	**2011**	**2012**	**2013**
12	Net Income after Taxes (Expansion)	NA		
13	End-of-year Cash on Hand (Expansion)	NA		
14	Net Income after Taxes (No Expansion)	NA		
15	End-of-year Cash on Hand (No Expansion)	NA		
16				
17	**Calculations (Expansion)**	**2011**	**2012**	**2013**
18	Total Sales Dollars	$350,000		
19	Cost of Goods Sold	$245,000		
20	Cost of Goods Sold (as a percent of Sales)	70%		
21	Interest Rate for Business Loan		NA	NA
22				
23	**Calculations (No Expansion)**	**2011**	**2012**	**2013**
24	Total Sales Dollars	$350,000		
25	Cost of Goods Sold	$245,000		
26	Cost of Goods Sold (as a percent of Sales)	70%		

FIGURE C-1 Tutorial skeleton 1

	A	B	C	D
28	**Income and Cash Flow Statements (Expansion)**	**2011**	**2012**	**2013**
29	Beginning-of-year Cash on Hand	NA		
30	Sales (Revenue)	NA		
31	Cost of Goods Sold	NA		
32	*Business Loan Payment*	NA		
33	Income before Taxes	NA		
34	Income Tax Expense	NA		
35	Net Income after Taxes	NA		
36	End-of-year Cash on Hand	$15,000		
37				
38	**Income and Cash Flow Statements (No Expansion)**	**2011**	**2012**	**2013**
39	Beginning-of-year Cash on Hand	NA		
40	Sales (Revenue)	NA		
41	Cost of Goods Sold	NA		
42	Income before Taxes	NA		
43	Income Tax Expense	NA		
44	Net Income after Taxes	NA		
45	End-of-year Cash on Hand	$15,000		

FIGURE C-2 Tutorial skeleton 2

Each spreadsheet section is discussed in detail next.

The Constants Section

This section holds values that are needed for the spreadsheet calculations. These values are usually given to you, and generally do not change for the exercise. However, you can change these values later if necessary; for example, you might need to borrow more or less money for your business expansion (cell C5). For this tutorial, the constants are the Tax Rate and the Loan Amount.

The Inputs Section

The Inputs section in Figure C-1 provides a place to designate the two possible economic outlooks and the two possible inflation outlooks. If you wanted to make these outlooks change by business year, you could leave blanks under both business years. However, as you will see later when you use Scenario Manager, this approach would greatly increase the complexity of interpreting the results. For simplicity's sake, assume that the same outlooks will apply to both years 2012 and 2013.

The Summary of Key Results Section

This section summarizes the Year 2 and 3 Net Income after Taxes (profit) and the End-of-year Cash on Hand both for expanding the business and for not expanding. These cells are copied from the Income and Cash Flow Statements section at the bottom of the sheet. Summary sections are frequently placed near the top of a spreadsheet to allow managers to see a quick "bottom line" summary without having to scroll down the spreadsheet to see the final result. Summary sections can also make it easier to select cells for charting.

The Calculations Sections (Expansion and No Expansion)

The following areas are used to compute the following necessary results:

- The Total Sales Dollars, which is a function of the Year 2011 value and the Economic Outlook input
- The Cost of Goods Sold, which is the Total Sales Dollars multiplied by the Cost of Goods Sold (as a percent of Sales)
- The Cost of Goods Sold (as a percent of Sales), which is a function of the Year 2011 value and the Inflation Outlook input
- In addition, the Calculations section for the expansion includes the interest rate, which is also a function of the Economic Outlook input. This interest rate will be used to determine the Business Loan Payment in the Income and Cash Flow Statements section.

You could make these formulas part of the Income and Cash Flow Statements section. However, it makes more sense to use the approach shown here because it makes the formulas in the Income and Cash Flow

Statements less complicated. In addition, when you create other DSS models that include unit costing and pricing calculations, you can enter the formulas in this section to facilitate managerial accounting cost analysis.

The Income and Cash Flow Statements Sections (Expansion and No Expansion)

These sections are the financial or accounting "body" of the spreadsheet. They contain the following values:

- Beginning-of-year Cash on Hand, which equals the *prior* year's End-of-year Cash on Hand.
- Sales (Revenue), which in this tutorial is simply the results of the Total Sales Dollars copied from the Calculations section.
- Cost of Goods Sold, which also is copied from the Calculations section.
- Business Loan Payment, which is calculated using the PMT (Payment) function and the inputs for loan amount and interest rate from the Constants and Calculations sections. Note that only the Income and Cash Flow Statement for Expansion includes a value for Business Loan Payment. If you do not expand, you do not need to borrow the money.
- Income before Taxes, which is Sales minus the Cost of Goods Sold; for the expansion scenarios, you also subtract the Business Loan Payment.
- Income Tax Expense, which is zero when there is no income or negative income; otherwise, this value is the Income before Taxes multiplied by the Tax Rate from the Constants section.
- Net Income after Taxes, which is Income before Taxes minus Income Tax Expense.
- End-of-year Cash on Hand, which is Beginning-of-year Cash on Hand plus Net Income after Taxes.

Note that this Income and Cash Flow Statement is greatly simplified. It does not address the issues of changes in Inventories, Accounts Payable, and Accounts Receivable, nor any period expenses such as Selling and General Administrative expenses, utilities, salaries, real estate taxes, insurance, or depreciation.

Construction of the Spreadsheet Model

Next, you will work through three steps to build the spreadsheet model:

1. Make a skeleton or "shell" of the spreadsheet. Save it with a name you can easily recognize, such as TUTC.xlsx or Tutorial C *YourName*.xlsx. When submitting electronic work to an instructor or supervisor, it is always a good idea to include your last name and first initial in the filename.
2. Fill in the "easy" cell formulas.
3. Then enter the "hard" spreadsheet formulas.

Making a Skeleton or "Shell"

The first step is to set up the skeleton worksheet. The skeleton should have headings, text labels, and constants. Do not enter any formulas yet.

Before you start entering data, you should first try to visualize a sensible structure for your worksheet. In Figures C-1 and C-2, the seven sections are arranged vertically down the page; the item descriptions are in the first column (A), and the time periods (years) are in the next three columns (B, C, and D). This is a widely accepted business practice, and is commonly called a "horizontal analysis." It is used to visually compare financial data side by side through successive time periods.

Because your key results depend on the Income and Cash Flow Statements, you usually set up that section first, and then work upward to the top of the sheet. In other words, you set up the Income and Cash Flow Statements section, then the Calculations section, and then the Summary of Key Results, Inputs, and Constants sections. Some might argue that the Income and Cash Flow Statements should be at the top of the sheet, but when you want to change values in the Constants or Inputs section or examine the Summary of Key Results, it does not make sense to have to scroll to the bottom of the worksheet. When you run the model, you do not enter anything in the Income and Cash Flow Statements—they are all calculations. So, it makes sense to put them last.

Here are some other general guidelines for designing effective DSS spreadsheets:

- Decide which items belong in the Calculations section. A good rule of thumb is that if your items have formulas but do not belong in the Income and Cash Flow Statements, put them in the Calculations section. Good examples are intermediate calculations such as unit volumes, costs and prices, markups, or changing interest rates.

- The Summary of Key Results section should be just that—*key* results. These outputs help you make good business decisions. Key results frequently include net income before taxes (profit) and end-of-year cash on hand (how much cash your business has). However, if you are creating a DSS model on alternative capital projects, your key results can also include cost savings, net present value of a project, or rate of return for an investment.
- The Constants section is for known values you need to perform other calculations. You use a Constants section rather than just including those values in formulas so that you can input new values if they change, such as tax rates. It makes your DSS model more flexible.

AT THE KEYBOARD

Enter the Excel skeleton shown in Figures C-1 and C-2.

NOTE

When you see NA (Not Applicable) in a cell, do not enter any values or formulas in the cell. The cells that contain values in the 2011 column are used by other cells for calculations. In this example, you are mainly interested in what happens in 2012 and 2013. The rest of the cells are "Not Applicable."

Filling in the "Easy" Formulas

The next step in building a spreadsheet is to fill in the "easy" formulas. To begin, format all the cells that will contain monetary values as Currency with zero decimal places:

- Constants—C5
- Summary of Key Results—C12 to C15, D12 to D15
- Calculations (Expansion)—C18, C19, D18, D19
- Calculations (No Expansion)—C24, C25, D24, D25
- Income and Cash Flow Statements (Expansion)—B36, C29 to C36, D29 to D36
- Income and Cash Flow Statements (No Expansion)—B45, C39 to C45, D39 to D45

NOTE

With the insertion point at cell C12 (where the $0 appears), note the editing window—the white space at the top of the spreadsheet to the right of the *fx* symbol. The cell's contents, whether it is a formula or value, should appear in the editing window. In this case, the window shows =C35.

The Summary of Key Results section (see Figure C-3) will contain the values you calculate in the Income and Cash Flow Statements sections. To copy the cell contents for this section, move your mouse cursor to cell C12, click the cell, type =C35, and press Enter. If you formatted your money cells properly, a $0 should appear in cell C12.

	C12		*fx*	=C35		
	A			B	C	D
11	**Summary of Key Results**			**2011**	**2012**	**2013**
12	Net Income after Taxes (Expansion)			NA	$0	
13	End-of-year Cash on Hand (Expansion)			NA		
14	Net Income after Taxes (No Expansion)			NA		
15	End-of-year Cash on Hand (No Expansion)			NA		

FIGURE C-3   Value from cell C35 (Net Income after Taxes) copied to cell C12

Because cell C35 does not contain a value yet, Excel assumes that the empty cell has a numerical value of 0. When you put a formula in cell C35 later, cell C12 will echo the resulting answer. Because Net Income after Taxes (Expansion) for 2013 (cell D35) and its corresponding cell in Summary of Key Results (cell D12) are both directly to the right of the values for 2012, you can either type =D35 into cell D12 or copy cell C12 to D12. To perform the copy operation:

1. Click in the cell or range of cells you want to copy.
2. Hold down the Control key and press C (Ctrl+C).
3. A marquee (a moving dashed box) should now be animated over the cells selected for copying.
4. Select your destination cell or range of cells.
5. Hold down the Control key and press V (Ctrl+V). Cell D12 should now contain $0, but actually it has a reference to cell D35. Click cell D12 and look again at the editing window; it should display =D35.

Cells C14, C15, D14, and D15 represent Net Income after Taxes and End-of-year Cash on Hand for both years of No Expansion; these cells are mirrors of cells C44, C45, D44, and D45 in the last section. Select cell C14, type =C44, and press Enter. Select cell C14 again, use the Copy command, and paste the contents into cell D14 (see Figure C-4).

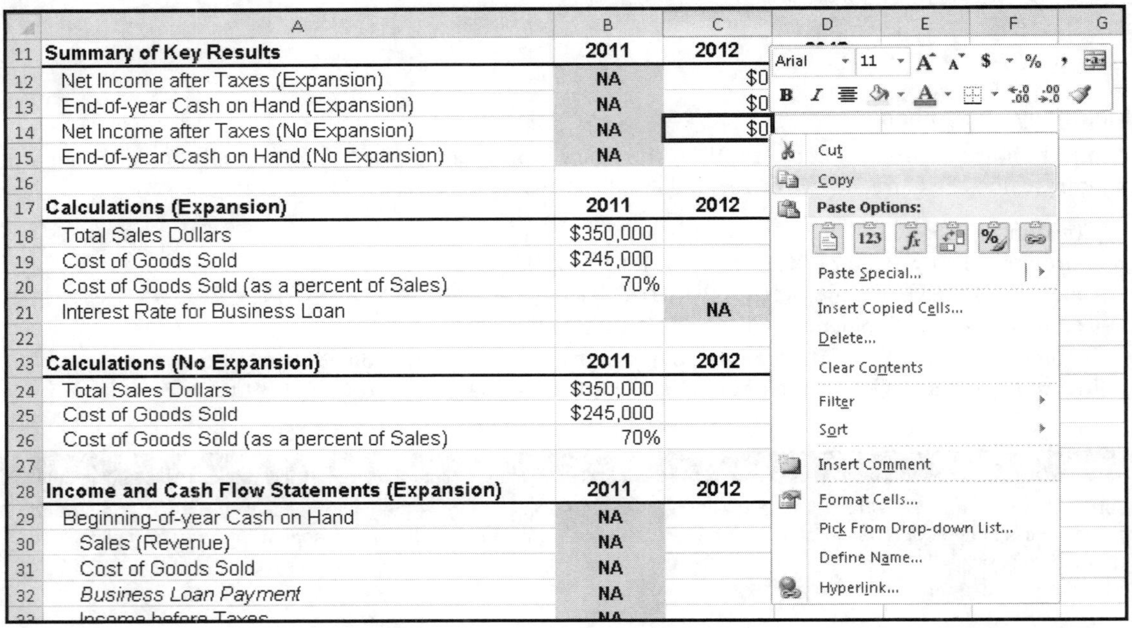

FIGURE C-4 Copying the formula from cell C14 to cell D14

Because Excel uses *relative* cell references by default, copying cell C14 into cell D14 will copy and paste the contents of cell D44 (the cell adjacent to C44) into cell D14. See Figure C-5.

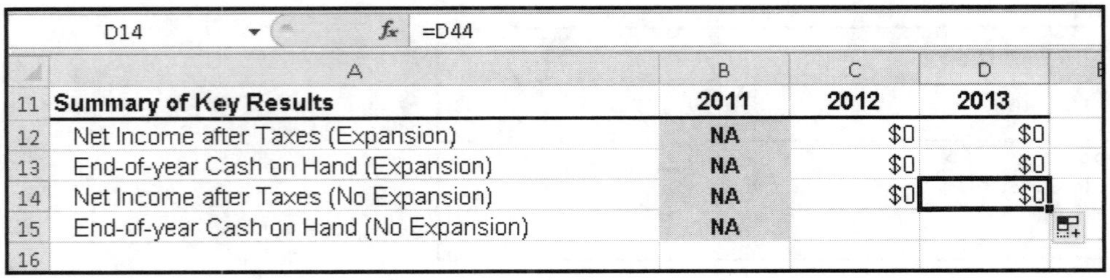

FIGURE C-5 Formula from cell D44 pasted into cell D14

Use the Copy command again, this time downward from cells C14 and D14, to complete cells C15 and D15. If you are successful, the formula in the editing window for cell C15 will be "=C45" and cell D15 will display "=D45".

You will create the formulas for the two Calculations sections last because they are the hardest formulas. Next, you will create the formulas for the two Income and Cash Flow Statements sections; all the cells in these two sections should be formatted as Currency with zero decimal places.

As shown in Figure C-6, the Beginning-of-year Cash on Hand for 2012 is the End-of-year Cash on Hand for 2011. In cell C29, type =B36. A handy shortcut is to type the "=" sign, immediately move your mouse pointer to the cell you want to designate, and then click the left mouse button. Excel will enter the cell location into the formula for you. This shortcut is especially useful if you want to avoid making a typing error.

	A	B	C	D
	SUM ▾ ✗ ✓ *fx* =B36			
28	**Income and Cash Flow Statements (Expansion)**	**2011**	**2012**	**2013**
29	Beginning-of-year Cash on Hand	NA	=B36	
30	Sales (Revenue)	NA		
31	Cost of Goods Sold	NA		
32	*Business Loan Payment*	NA		
33	Income before Taxes	NA		
34	Income Tax Expense	NA		
35	Net Income after Taxes	NA		
36	End-of-year Cash on Hand	$15,000		
37				
38	**Income and Cash Flow Statements (No Expansion)**	**2011**	**2012**	**2013**
39	Beginning-of-year Cash on Hand	NA		
40	Sales (Revenue)	NA		
41	Cost of Goods Sold	NA		
42	Income before Taxes	NA		
43	Income Tax Expense	NA		
44	Net Income after Taxes	NA		
45	End-of-year Cash on Hand	$15,000		

FIGURE C-6   End-of-year Cash on Hand for 2011 copied to Beginning-of-year Cash on Hand for 2012

Likewise, copy the other three End-of-year Cash on Hand cells to the Beginning-of-year Cash on Hand cells for both Income and Cash Flow Statements (cells D29, C39, and D39).

The Sales (Revenue) cells C30, D30, C40, and D40 are simply copies of the contents of cells C18, D18, C24, and D24, respectively, from the Calculations sections (both Expansion and No Expansion). Use the shortcut method to copy these cells. Note that all four cells will display $0 until you enter the formulas in the Calculations sections (see Figure C-7).

	A	B	C	D
	D40 ▾ *fx* =D24			
28	**Income and Cash Flow Statements (Expansion)**	**2011**	**2012**	**2013**
29	Beginning-of-year Cash on Hand	NA	$15,000	$0
30	Sales (Revenue)	NA	$0	$0
31	Cost of Goods Sold	NA		
32	*Business Loan Payment*	NA		
33	Income before Taxes	NA		
34	Income Tax Expense	NA		
35	Net Income after Taxes	NA		
36	End-of-year Cash on Hand	$15,000		
37				
38	**Income and Cash Flow Statements (No Expansion)**	**2011**	**2012**	**2013**
39	Beginning-of-year Cash on Hand	NA	$15,000	$0
40	Sales (Revenue)	NA	$0	$0
41	Cost of Goods Sold	NA		
42	Income before Taxes	NA		
43	Income Tax Expense	NA		
44	Net Income after Taxes	NA		
45	End-of-year Cash on Hand	$15,000		

FIGURE C-7   Sales Revenue cells copied from the Calculations sections

The Cost of Goods Sold cells C31, D31, C41, and D41 are simply copies of the contents of cells C19, D19, C25, and D25, respectively, from the Calculations sections. Because the cells in both locations are directly below the Sales cells in the four locations, you can use the Copy command to fill those cells easily. As you can see in Figure C-8, you can drag your mouse pointer over both cells C40 and D40, right-click to see the floating toolbar, and select Copy. Move your mouse pointer to select cells C41 and D41, right-click the mouse, and select Paste. If you are uncomfortable working with the mouse, you can type =C19, =D19, =C25, and =D25 in cells C31, D31, C41, and D41.

	A	B	C	D	E	F	G	H
28	**Income and Cash Flow Statements (Expansion)**	**2011**	**2012**	**2013**				
29	Beginning-of-year Cash on Hand	NA	$15,000	$0				
30	Sales (Revenue)	NA	$0	$0				
31	Cost of Goods Sold	NA	$0	$0				
32	*Business Loan Payment*	NA						
33	Income before Taxes	NA						
34	Income Tax Expense	NA						
35	Net Income after Taxes	NA						
36	End-of-year Cash on Hand	$15,000						
37								
38	**Income and Cash Flow Statements (No Expansion)**	**2011**	**2012**	**2013**				
39	Beginning-of-year Cash on Hand	NA	$15,000	$0				
40	Sales (Revenue)	NA	$0	$0				
41	Cost of Goods Sold	NA						
42	Income before Taxes	NA						
43	Income Tax Expense	NA						
44	Net Income after Taxes	NA						
45	End-of-year Cash on Hand	$15,000						

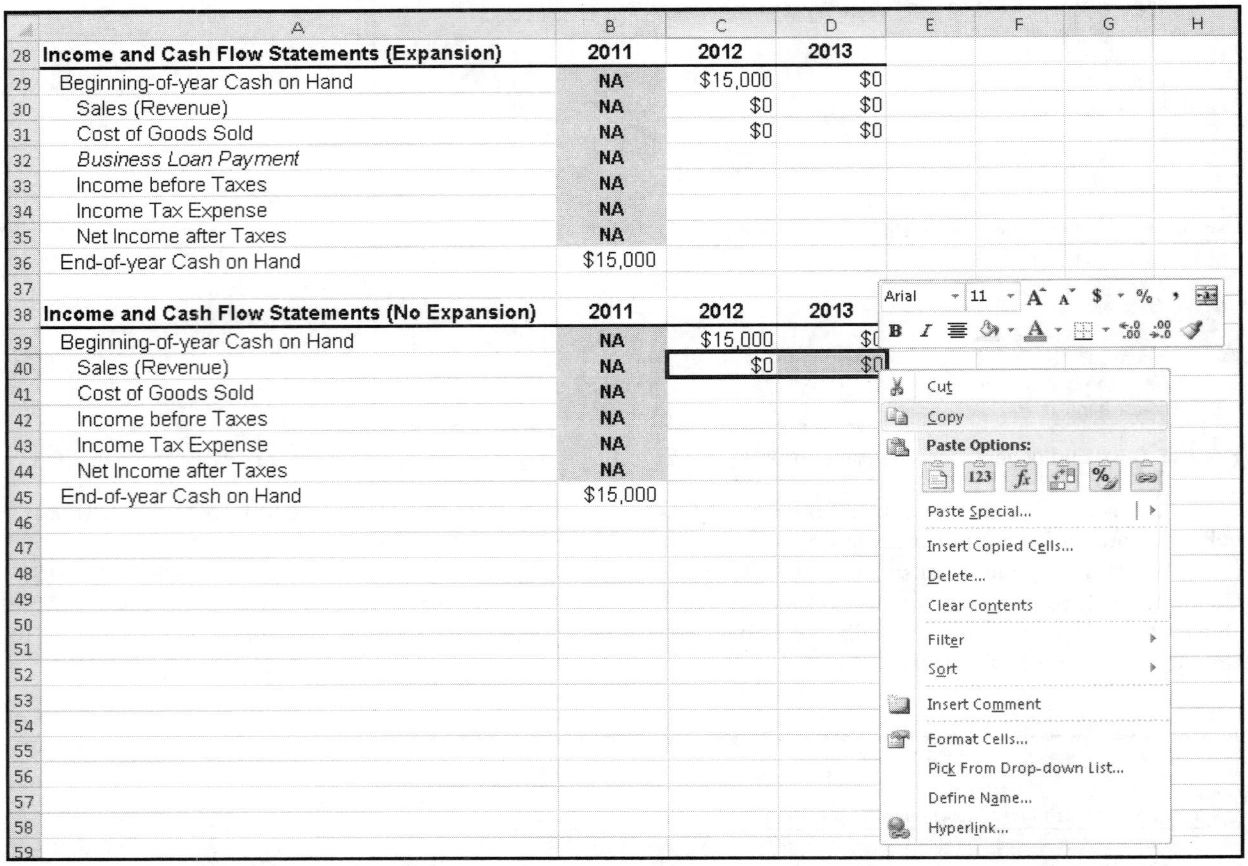

Floating toolbar options shown: Arial, 11, with formatting buttons. Right-click menu: Cut, Copy, Paste Options, Paste Special..., Insert Copied Cells..., Delete..., Clear Contents, Filter, Sort, Insert Comment, Format Cells..., Pick From Drop-down List..., Define Name..., Hyperlink...

FIGURE C-8   Cost of Goods Sold cells copied from the Calculations sections

Next you determine the Business Loan Payment for cells C32 and D32—notice that it is only present in the Income and Cash Flow Statements (Expansion) section, because if you do not expand the business, you do not need the business loan of $100,000. Excel has financial formulas to figure out loan payments. To determine a loan payment, you need to know three things: the amount being borrowed (cell C5 in the Constants section), the interest rate (cell B21 in the Calculations-Expansion section), and the number of payment periods. At the beginning of the tutorial, you learned that the bank was willing to loan money at either 5% or 4% interest compounded annually, to be paid over 10 years. Normally, banks require businesses to make monthly payments on their loans and compound the interest monthly, in which case you would enter 120 (12 months/year × 10 years) for the number of payments and divide the annual interest rate by 12 for the period interest rate. This formula is important to remember when you enter the business world, but for now you will simplify the calculation by specifying one loan payment per year compounded annually. To put in the payment formula, click cell C32, then click the *fx* symbol next to the editing window (circled in Figure C-9). The Payment function is called PMT, so type PMT in the Insert Function window—you will immediately see a short description of the function with its arguments, as shown in Figure C-9.

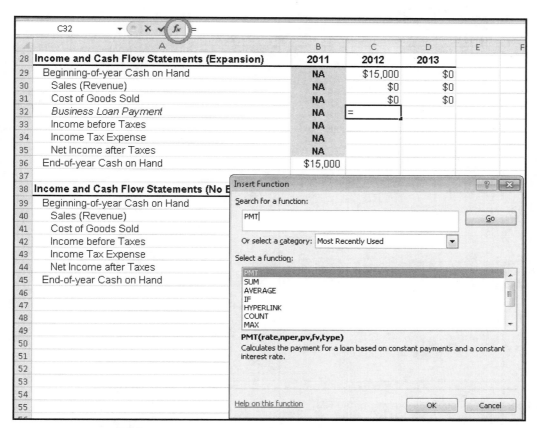

FIGURE C-9 Accessing the PMT function in Excel for cell C32

NOTE

Rate is the interest rate per period of the loan, Nper is an abbreviation for the number of loan periods, and Pv is an abbreviation for Present Value, the amount of money you are borrowing "today." The PMT function can determine a series of equal loan payments necessary to pay back the amount borrowed, plus the accumulated compound interest over the life of the loan.

When you click OK, the resulting window allows you to enter the cells or values needed in the function arguments (see Figure C-10). In the Rate text box, enter B21, which is the cell that will contain the calculated interest rate. In the Nper text box, enter 10 (for 10 years). In the Pv text box, enter C5, which is the cell that contains the loan amount.

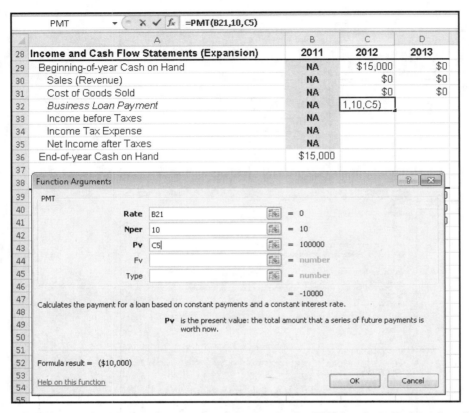

FIGURE C-10 The Function Arguments dialog box for the PMT function with the values filled in

When you click OK, ($10,000) should appear in cell C32. Payments in Excel always appear as negative numbers, which is why the number has parentheses around it. (Depending on your cell formatting, the number may also appear in red.) Next, you need to have the same payment amount in cell D32 (for 2013). Because the PMT function creates equal payments over the life of the loan, you can simply type =C32 into cell D32.

The next line in the Income and Cash Flow Statements is Income before Taxes, which is an easy calculation. It is the Sales minus the Cost of Goods Sold, minus the Business Loan Payment. However, because the PMT function shows the loan payment as a negative number, you will instead add the Business Loan Payment. In cell C33, enter =C30-C31+C32. Again, a negative $10,000 should be displayed, as the cells other than the loan payment currently have zero in them. Copy cell C33 to cell D33. In cell C42 of the next section below (No Expansion), enter =C40-C41. (There is no loan payment in this section to put in the calculation.) Next, copy cell C42 to cell D42. At this point, your Income and Cash Flow Statements should look like Figure C-11.

D42	fx	=D40-D41		
	A	B	C	D
28 **Income and Cash Flow Statements (Expansion)**		**2011**	**2012**	**2013**
29	Beginning-of-year Cash on Hand	NA	$15,000	$0
30	Sales (Revenue)	NA	$0	$0
31	Cost of Goods Sold	NA	$0	$0
32	*Business Loan Payment*	NA	($10,000)	($10,000)
33	Income before Taxes	NA	-$10,000	-$10,000
34	Income Tax Expense	NA		
35	Net Income after Taxes	NA		
36	End-of-year Cash on Hand	$15,000		
37				
38 **Income and Cash Flow Statements (No Expansion)**		**2011**	**2012**	**2013**
39	Beginning-of-year Cash on Hand	NA	$15,000	$0
40	Sales (Revenue)	NA	$0	$0
41	Cost of Goods Sold	NA	$0	$0
42	Income before Taxes	NA	$0	$0
43	Income Tax Expense	NA		
44	Net Income after Taxes	NA		
45	End-of-year Cash on Hand	$15,000		

FIGURE C-11   The Income and Cash Flow Statements completed up to Income before Taxes

Income Tax Expense is the most complex formula for these sections. Because you do not pay income tax when you have no income or a loss, you must use a formula that allows you to enter 0 if there is no income or a loss, or to calculate the tax rate on a positive income. The IF function in Excel allows you to enter one of two different results in a cell, depending on whether a defined logical statement is true or false. To create an IF function, select cell C34, then click the *fx* symbol next to the cell editing window (circled in Figure C-12). When the Insert Function dialog box appears, type IF in the "Search for a function" text box, and click the Go button if necessary. The IF function should appear. When you click OK, the Function Arguments dialog box appears (see Figure C-13).

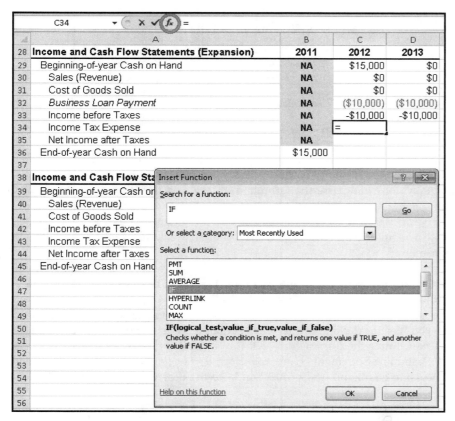

FIGURE C-12   The IF function

Type the following in the Function Arguments dialog box:

- Next to Logical_test, type C33<=0.
- Next to Value_if_true, type 0.
- Next to Value_if_false, type C33*C4 (the Income before Taxes multiplied by the Tax Rate for 2012).

As you fill in the arguments, Excel writes the formula for you in the formula editing window (circled in Figure C-13).

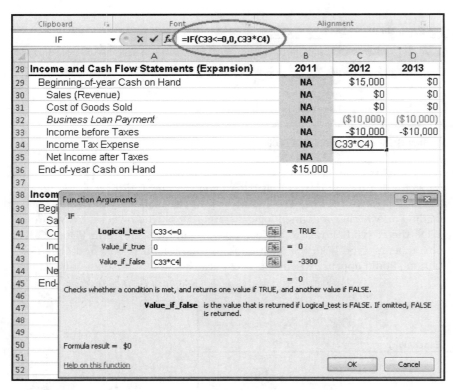

FIGURE C-13   The Function Arguments dialog box with the arguments filled in

Once you have typed in the arguments, click OK; Excel enters the formula into the cell. Because you had negative income, the cell should display a zero for now. Because the same formula will be used in 2013 (but with the 2013 tax rate), you can simply copy and paste the formula from cell C34 to cell D34. You also have to calculate the income tax for the Income and Cash Flow Statements (No Expansion). In cell C43, use the same IF function, but in the Logical_test, Value_if_true, and Value_if_false arguments, you must type C42<=0, 0, and C42*C4, respectively. Again, the cell will display $0 for an answer. Copy cell C43 to cell D43 to complete the Income Tax Expense line.

Net Income after Taxes is simply the Income before Taxes minus the Income Tax Expense. Enter the formula into cell C35, then copy cell C35 over to cells D35, C44, and D44. If you did this correctly, cells C35 and D35 will display a negative $10,000, and cells C44 and D44 will display $0.

End-of-year Cash on Hand, the last line in both Income and Cash Flow Statements sections, is not diffi-cult either. Conceptually, the cash you have at the end of the year is equal to your Beginning-of-year Cash on Hand plus your Net Income after Taxes. Enter the formula into cell C36, then copy cell C36 over to cell D36. Note that because the Income and Cash Flow Statements (No Expansion) do not have a line item for Business Loan Payment, you cannot copy the same command down to it. You have to enter the formula manually for cell C45, which is =C39+C44. However, you can copy cell C45 to cell D45 to finish the Income and Cash Flow Statements sections. The completed sections should look like Figure C-14.

D45		f_x =D39+D44		
	A	B	C	D
28	**Income and Cash Flow Statements (Expansion)**	**2011**	**2012**	**2013**
29	Beginning-of-year Cash on Hand	NA	$15,000	$5,000
30	Sales (Revenue)	NA	$0	$0
31	Cost of Goods Sold	NA	$0	$0
32	*Business Loan Payment*	NA	($10,000)	($10,000)
33	Income before Taxes	NA	-$10,000	-$10,000
34	Income Tax Expense	NA	$0	$0
35	Net Income after Taxes	NA	-$10,000	-$10,000
36	End-of-year Cash on Hand	$15,000	$5,000	-$5,000
37				
38	**Income and Cash Flow Statements (No Expansion)**	**2011**	**2012**	**2013**
39	Beginning-of-year Cash on Hand	NA	$15,000	$15,000
40	Sales (Revenue)	NA	$0	$0
41	Cost of Goods Sold	NA	$0	$0
42	Income before Taxes	NA	$0	$0
43	Income Tax Expense	NA	$0	$0
44	Net Income after Taxes	NA	$0	$0
45	End-of-year Cash on Hand	$15,000	$15,000	$15,000

FIGURE C-14   The completed Income and Cash Flow Statements sections

Filling in the "Hard" Formulas

To finish the spreadsheet, you will enter values in the Inputs section and write the formulas in both Calculations sections.

AT THE KEYBOARD

In cell C8, enter an R for Recession, and in cell C9, enter H for High Inflation. You could enter any values here, but these two values will work with the IF functions you will write later. Recall that you did not use separate inputs for 2012 and 2013. You are assuming that the economic outlook or inflation rate that exists for 2012 will extend into 2013. However, because you are using the same inputs from these two locations, you must remember to use *absolute* cell references to both cells C8 and C9 in the various IF statements if you want to use a Copy command for adjacent cells. Your Inputs section should look like the one in Figure C-15.

	A	B	C	D
7	**Inputs**	**2011**	**2012**	**2013**
8	Economic Outlook (R=Recession, B=Boom)	NA	R	NA
9	Inflation Outlook (H=High, L=Low)	NA	H	NA

FIGURE C-15   The Inputs section with values entered in cells C8 and C9

Remember that you referred to cell addresses in both Calculations sections in your formulas in the Income and Cash Flow Statements sections. Now you will enter formulas for these calculations. If necessary, format the four Total Sales Dollars cells and the four Cost of Goods Sold cells in the Calculations sections as Currency with no decimal places.

As described at the beginning of the tutorial, the forecast for Total Sales Dollars is a function of both the Economic Outlook and whether you expand the business. The following table lists the predicted sales growth percentages:

Sales Growth Forecast—Collegetown Thrift Shop

	Business Expansion	No Business Expansion
Recession-R	30%	20%
Boom-B	15%	5%

You will use IF formulas to forecast Total Sales Dollars. Click cell C18, then bring up the IF function and type the following in the text boxes:

Logical_test: $C$8="R" (Note that you must use absolute cell referencing for B8 plus quotation marks for Excel to recognize a text string.)

Value_if_true: B18*1.3 (the 2011 sales multiplied by 1.3 for 30% sales growth)

Value_if_false: B18*1.15 (the 2011 sales multiplied by 1.15 for 15% sales growth)

Compare your entries to Figure C-16.

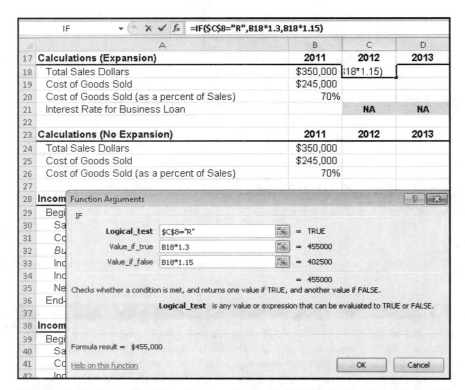

FIGURE C-16   Using the IF function to enter the Total Sales Dollars forecast for 2012

When you click OK, cell C18 should display $455,000, because 30% of $350,000 is $105,000, and $350,000 plus $105,000 equals $455,000. So, it appears that the formula returned a "true" value with an R inserted in cell C8. Because you "anchored" cell C8 by entering $C$8, copy this formula over to cell D18 for the year 2013.

Once you complete the Total Sales Dollars cells for the Expansion scenario, go down to the Calculations (No Expansion) section and use IF statements to enter formulas for the Total Sales Dollars. Use 20% for Recession and 5% for Boom. You can copy the formula from cell C18 into cell C24, but you then will have to use the editing window to change the values in the true and false arguments from 1.3 and 1.15 to 1.2 and 1.05, respectively, to reflect the fact that you did not expand the business. See Figure C-17.

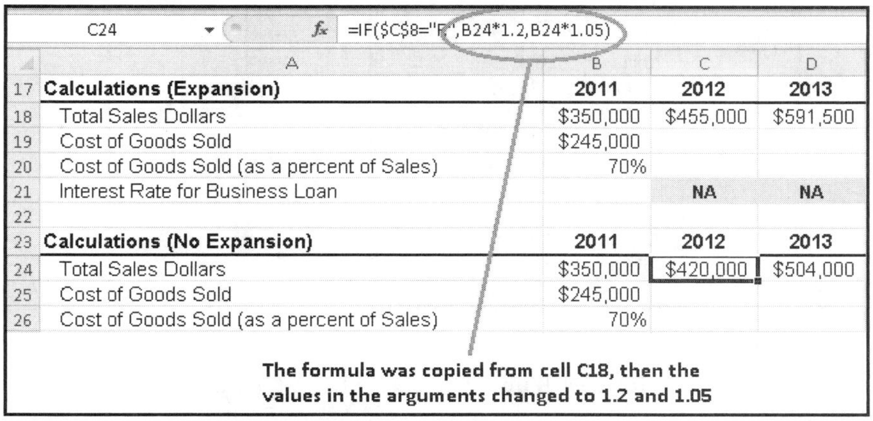

C24		f_x =IF($C$8="F",B24*1.2,B24*1.05)			
	A	B	C	D	
17	**Calculations (Expansion)**	**2011**	**2012**	**2013**	
18	Total Sales Dollars	$350,000	$455,000	$591,500	
19	Cost of Goods Sold	$245,000			
20	Cost of Goods Sold (as a percent of Sales)	70%			
21	Interest Rate for Business Loan		NA	NA	
22					
23	**Calculations (No Expansion)**	**2011**	**2012**	**2013**	
24	Total Sales Dollars	$350,000	$420,000	$504,000	
25	Cost of Goods Sold	$245,000			
26	Cost of Goods Sold (as a percent of Sales)	70%			

The formula was copied from cell C18, then the
values in the arguments changed to 1.2 and 1.05

FIGURE C-17   Copying cell C18 into cell C24 and then editing the IF function arguments to change the sales growth percentages

As before, you can now copy cell C24 to cell D24. You have completed the Total Sales Dollars calculations.

The Cost of Goods Sold (cells C19, D19, C25, and D25) is the Total Sales Dollars multiplied by the Cost of Goods Sold as a percent of Sales. In cell C19, type =C18*C20 and press Enter. Copy cell C19 and paste the contents into cells D19, C25, and D25. Your answers will be $0 until you enter the formulas for the Cost of Goods Sold as a percent of Sales.

Cost of Goods Sold as a percent of Sales (cells C20, D20, C26, and D26) was 70% in 2011. In variety merchandising for resold items, it is easier to use an aggregate measure such as Cost of Goods Sold as a percent of Sales rather than trying to capture an individual Cost of Goods Sold for each item. From the 2011 data, you determined that for every dollar of sales you collected in 2011, you spent 70 cents purchasing the item for resale. You will use that percentage as a basis for forecasting Cost of Goods Sold as a percent of Sales, applying an appropriate inflation factor for the cost of acquiring the stock for sale. The following table lists the predicted inflation percentages for Cost of Goods Sold.

Cost of Goods Sold Forecast—Collegetown Thrift Shop

	Business Expansion	No Business Expansion
High Inflation	6%	6%
Low Inflation	2%	2%

As with Total Sales Dollars previously, you will again use the IF function to calculate the Cost of Goods Sold as a percent of Sales. Now that you are familiar with the IF function, you can probably enter the function without using the dialog boxes. In cell C20, type the following:

=IF($C$9="H",B20*1.06,B20*1.02)

This expression means that if the text string in cell C9 is the letter H, you multiply the value in cell B20 by 1.06 (6% inflation). If the value in cell C9 is not an H, multiply the value in cell B20 by 1.02 (2% inflation). The value in cell B20 was the baseline Cost of Goods Sold as a percent of Sales in 2011, which was 70%. You can now copy cell C20 and paste the contents into cell D20.

Because the inflation percentages were exactly the same for both the Expansion and No Expansion calculations, you can also copy cell C20 and paste the contents into cells C26 and D26. Your Calculations sections should now look like Figure C-18.

D26	▼	fx	=IF($C$9="H",C26*1.06,C26*1.02)		

	A	B	C	D
17	**Calculations (Expansion)**	2011	2012	2013
18	Total Sales Dollars	$350,000	$455,000	$591,500
19	Cost of Goods Sold	$245,000	$337,610	$465,227
20	Cost of Goods Sold (as a percent of Sales)	70%	74%	79%
21	Interest Rate for Business Loan		NA	NA
22				
23	**Calculations (No Expansion)**	2011	2012	2013
24	Total Sales Dollars	$350,000	$420,000	$504,000
25	Cost of Goods Sold	$245,000	$311,640	$396,406
26	Cost of Goods Sold (as a percent of Sales)	70%	74%	79%

FIGURE C-18   Calculations sections nearly complete

The last item in the Calculations section is the Interest Rate for Business Loan (cell B21). Remember the bank's statement that if the economy recovers, it could lower the interest rate from 5% to 4%. So, you will need one more IF function to insert into cell B21 based on the economic outlook. If the economic outlook is for a Recession (R), then the interest rate will be 5% annually; if the outlook is for a Boom (B), then the interest rate will be 4% annually. Now that you are familiar with the IF function, you can simply type the expression into the cell yourself. Click cell B21, type =IF(C8="R",5%,4%), and press Enter.

You will immediately notice that 5% appears in the cell because you have R in the input cell for Economic Outlook. You may also notice that you now have a negative $12,950 in the Business Loan Payment cells (C32 and D32). See Figure C-19 to compare your results.

	A	B	C	D
7	**Inputs**	2011	2012	2013
8	Economic Outlook (R=Recession, B=Boom)	NA	R	NA
9	Inflation Outlook (H=High, L=Low)	NA	H	NA
10				
11	**Summary of Key Results**	2011	2012	2013
12	Net Income after Taxes (Expansion)	NA	$69,974	$73,660
13	End-of-year Cash on Hand (Expansion)	NA	$84,974	$158,634
14	Net Income after Taxes (No Expansion)	NA	$72,601	$69,936
15	End-of-year Cash on Hand (No Expansion)	NA	$87,601	$157,537
16				
17	**Calculations (Expansion)**	2011	2012	2013
18	Total Sales Dollars	$350,000	$455,000	$591,500
19	Cost of Goods Sold	$245,000	$337,610	$465,227
20	Cost of Goods Sold (as a percent of Sales)	70%	74%	79%
21	Interest Rate for Business Loan	5%	NA	NA
22				
23	**Calculations (No Expansion)**	2011	2012	2013
24	Total Sales Dollars	$350,000	$420,000	$504,000
25	Cost of Goods Sold	$245,000	$311,640	$396,406
26	Cost of Goods Sold (as a percent of Sales)	70%	74%	79%
27				
28	**Income and Cash Flow Statements (Expansion)**	2011	2012	2013
29	Beginning-of-year Cash on Hand	NA	$15,000	$84,974
30	Sales (Revenue)	NA	$455,000	$591,500
31	Cost of Goods Sold	NA	$337,610	$465,227
32	*Business Loan Payment*	NA	($12,950)	($12,950)
33	Income before Taxes	NA	$104,440	$113,323
34	Income Tax Expense	NA	$34,465	$39,663
35	Net Income after Taxes	NA	$69,974	$73,660
36	End-of-year Cash on Hand	$15,000	$84,974	$158,634
37				
38	**Income and Cash Flow Statements (No Expansion)**	2011	2012	2013
39	Beginning-of-year Cash on Hand	NA	$15,000	$87,601
40	Sales (Revenue)	NA	$420,000	$504,000
41	Cost of Goods Sold	NA	$311,640	$396,406
42	Income before Taxes	NA	$108,360	$107,594
43	Income Tax Expense	NA	$35,759	$37,658
44	Net Income after Taxes	NA	$72,601	$69,936
45	End-of-year Cash on Hand	$15,000	$87,601	$157,537

FIGURE C-19   The finished spreadsheet

You can change the economic inputs in four different combinations, R-H, R-L, B-H, B-L, and see the impact on your net income and cash on hand both for expanding and not expanding. However, you have another more powerful way to do this. In the next section, you will learn how to tabulate the financial results of the four possible combinations using an Excel tool called Scenario Manager.

SCENARIO MANAGER

You are now ready to evaluate the four possible outcomes for your DSS model. Because this is a simple, four-outcome model, you could have created four different spreadsheets, one for each set of outcomes, and then transferred the financial information from each spreadsheet to a Summary Report.

In essence, Scenario Manager performs exactly the same task. It runs the model for all the requested outcomes and presents a tabular summary of the results. This summary is especially useful for reports and presentations needed by upper managers, financial investors, or in this case, the bank.

To review, the four possible combinations of input values are: R-H (Recession and High Inflation), R-L (Recession and Low Inflation), B-H (Boom and High Inflation), and B-L (Boom and Low Inflation). You could consider each combination of inputs a separate scenario. For each of these scenarios, you are interested in four outputs: Net Income after Taxes for Expansion and No Expansion, and End-of-year Cash on Hand for Expansion and No Expansion.

Scenario Manager runs each set of combinations and then records the specified outputs as a summary into a separate worksheet. You can use these summary values as a table of numbers and print it, or you can copy them into a Microsoft Word document or a PowerPoint presentation. You can also use the data table to build a chart or graph, which you can put into a report or presentation.

When you define a scenario in Scenario Manager, you name it and identify the input cells and input values. Then you identify the output cells so Scenario Manager can capture the outputs in a summary sheet.

AT THE KEYBOARD

To start, select the Data tab on the Ribbon. In the Data Tools group, click the What-If Analysis icon, then click Scenario Manager from the menu that appears (see Figure C-20).

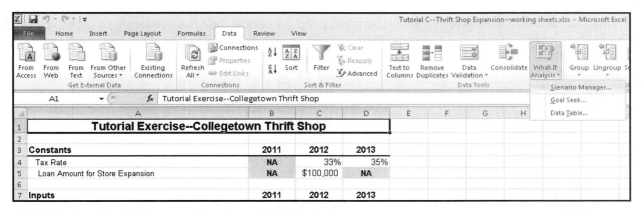

FIGURE C-20   Scenario Manager option in the What-If Analysis menu

Scenario Manager appears (see Figure C-21), but no scenarios are defined. Use the dialog box to add, delete, or edit scenarios.

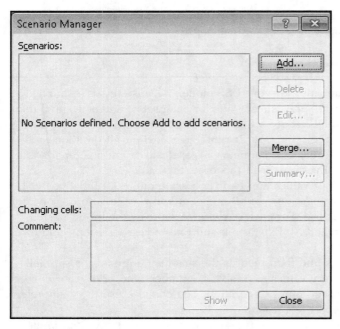

FIGURE C-21   Initial Scenario Manager dialog box

NOTE

When working with the Scenario Manager dialog box and any following dialog boxes, do not use the Enter key to navigate. Use mouse clicks to move from one step to the next.

To define a scenario, click the Add button. The Edit Scenario dialog box appears. Type Recession-High Inflation in the field under Scenario name. Then type the input cells in the Changing cells field (in this case, C8:C9). Better yet, you can use the button next to the field to select the cells in your spreadsheet. If you do, Scenario Manager changes the cell references to absolute cell references, which is acceptable (see Figure C-22).

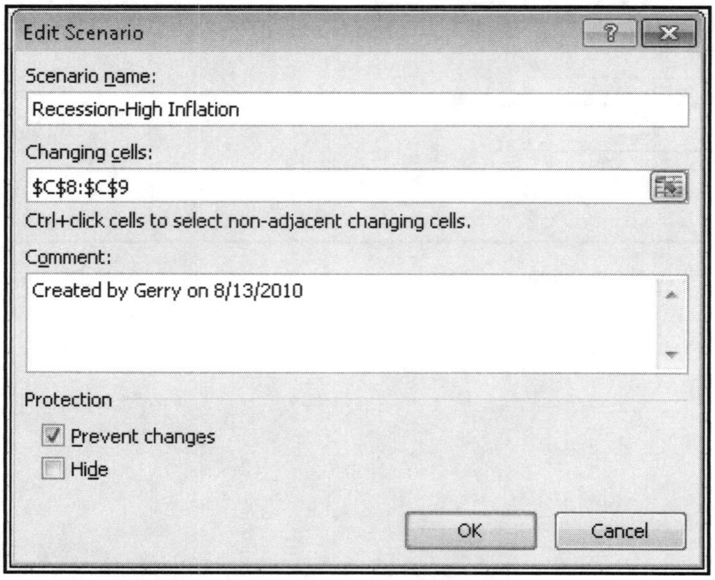

FIGURE C-22   Defining a scenario name and input cells

Click OK to open the Scenario Values dialog box. Enter the input values for the scenario. In the case of Recession and High Inflation, the values will be R and H for cells C8 and C9, respectively (see Figure C-23). Note that if you already have entered values in the spreadsheet, the dialog box will display the current values. Make sure to enter the correct values.

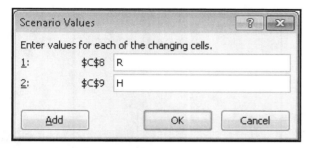

FIGURE C-23   Entering values for the input cells

Click OK to return to the Scenario Manager dialog box. Enter the other three scenarios: Recession-Low Inflation, Boom-High Inflation, and Boom-Low Inflation, and their related input values. When you finish, you should see the names and changing cells for the four scenarios (see Figure C-24).

FIGURE C-24   Scenario Manager dialog box with all four scenarios entered

You can now create a summary sheet that displays the results of running the four scenarios. Click the Summary button to open the Scenario Summary dialog box, as shown in Figure C-25. You must now enter the output cell addresses in Excel—they will be the same for all four scenarios. Recall that you created a section in your spreadsheet called Summary of Key Results. You are primarily interested in the results at the end of 2013, so you will choose the four cells that represent the Net Income after Taxes and End-of-year Cash on Hand, and then use them for both the expansion scenario and the non-expansion scenario. These cells are D12 to D15 in your spreadsheet. Either type in D12:D15 or use the button next to the Changing cells field and select those cells in the spreadsheet.

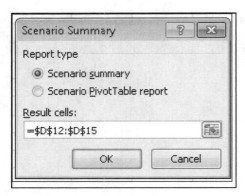

FIGURE C-25   Scenario Summary dialog box with Result cells entered

Another good reason for having a Summary of Key Results section is that it provides a contiguous range of cells to define for summary output. However, if you want to add output from other cells in the spreadsheet, simply separate each cell or range of cells in the dialog box with a comma. Next, click OK. Excel runs each set of inputs in the background, collects the results from the result cells, and then creates a new sheet called Scenario Summary (the name on the sheet's lower tab), as shown in Figure C-26.

		Current Values	Recession-High Inflation	Recession-Low Inflation	Boom-High Inflation	Boom-Low Inflation
Scenario Summary						
Changing Cells:						
$C$8	R	R	R	B	B	
$C$9	H	H	L	H	L	
Result Cells:						
$D$12		$73,660	$73,660	$96,052	$56,216	$73,738
$D$13		$158,634	$158,634	$189,562	$132,531	$157,605
$D$14		$69,936	$69,936	$89,015	$53,545	$68,152
$D$15		$157,537	$157,537	$184,496	$132,071	$153,573

Notes: Current Values column represents values of changing cells at time Scenario Summary Report was created. Changing cells for each scenario are highlighted in gray.

FIGURE C-26   Scenario Summary sheet created by Scenario Manager

As you can see, the output created by the Scenario Summary sheet is not formatted for easy reading. You do not know which results are the net income and cash on hand, and you do not know which results are for Expansion vs. No Expansion because Scenario Manager only listed the cell addresses. Scenario Manager also listed a separate column (column D) for the current input values in the spreadsheet, which are the same as the values in column E. It also left a blank column (column A) in the spreadsheet.

Fortunately, it is fairly easy to format the output. Delete columns D and A, put in the labels for cell addresses in the new column A, and then retitle the Scenario Summary as Collegetown Thrift Shop Financial Forecast, End of Year 2013 (because you are only looking at Year 2013 results). You can also make the results columns narrower by breaking the column headings into two lines; place your cursor in the editing window where you want to break the words, and then press Alt+Enter. Add a heading for column B (Cell Address). Finally, merge and center the title, and center the column headings and the input cell values (R, B, H, and L). Leave your financial data right-justified to keep the numbers lined up correctly. Finally, put some border boxes around each column of results. Figure C-27 shows a formatted Scenario Summary worksheet.

	A	B	C	D	E	F
1						
2	Scenario Summary--Collegetown Thrift Shop Financial Forecast, End of Year 2013					
3		Cell Address	Recession- High Inflation	Recession- Low Inflation	Boom- High Inflation	Boom- Low Inflation
5	Changing Cells:					
6	Economic Outlook: R-Recession, B-Boom	$C$8	R	R	B	B
7	Inflation: H-High, L-Low	$C$9	H	L	H	L
8	Result Cells:					
9	Net Income after Taxes (Expansion)	$D$12	$73,660	$96,052	$56,216	$73,738
10	End-of-year Cash on Hand (Expansion)	$D$13	$158,634	$189,562	$132,531	$157,605
11	Net Income after Taxes (No Expansion)	$D$14	$69,936	$89,015	$53,545	$68,152
12	End-of-year Cash on Hand (No Expansion)	$D$15	$157,537	$184,496	$132,071	$153,573
13	Notes: Current Values column represents values of changing cells at					
14	time Scenario Summary Report was created. Changing cells for each					
15	scenario are highlighted in gray.					

FIGURE C-27   Scenario Summary worksheet after formatting

Interpreting the Results

Now that you have good data, what do you do with it? Remember, you wanted to see if taking a $100,000 business loan to expand the thrift shop was a good financial decision. This is a relatively simple business case, and the shop's success so far ($350,000 of sales in 2011) would seem to make expansion a good risk. But how good?

After building the spreadsheet and doing the analysis, you can make comparisons and interpret the results. Regardless of the economic outlook or inflation, all four scenarios indicate that expanding the business should provide greater Net Income After Taxes and End-of-year Cash on Hand (after two years) than not expanding. So, the DSS model not only provides a quantitative basis for expanding, it provides an analysis that you can present to prospective lenders.

What decision would you make about expansion if you only looked at the 2012 forecast? You could go back to the original spreadsheet and look at the figures for 2012, or you can go to Scenario Manager and create a new summary, specifying the 2012 cells C12 through C15. See Figure C-28.

	A	B	C	D	E	F	G	H
7	Inputs	2011	2012	2013				
8	Economic Outlook (R=Recession, B=Boom)	NA	R	NA				
9	Inflation Outlook (H=High, L=Low)	NA	H	NA				
10								
11	Summary of Key Results	2011	2012	2013				
12	Net Income after Taxes (Expansion)	NA	$69,974	$73,660	Scenario Summary			
13	End-of-year Cash on Hand (Expansion)	NA	$84,974	$158,634	Report type			
14	Net Income after Taxes (No Expansion)	NA	$72,601	$69,936	● Scenario summary			
15	End-of-year Cash on Hand (No Expansion)	NA	$87,601	$157,537	○ Scenario PivotTable report			
16					Result cells:			
17	Calculations (Expansion)	2011	2012	2013	=$C$12:$C$15			
18	Total Sales Dollars	$350,000	$455,000	$591,500	OK Cancel			
19	Cost of Goods Sold	$245,000	$337,610	$465,227				
20	Cost of Goods Sold (as a percent of Sales)	70%	74%	79%				

FIGURE C-28   Creating a new Scenario Summary for 2012 instead of 2013

When you click OK, Excel creates a second Scenario Summary, but this time the output values come from 2012, not 2013. After editing and formatting, the 2012 Scenario Summary should look like Figure C-29.

	A	B	C	D	E	F
1						
2	Scenario Summary 2--Collegetown Thrift Shop Financial Forecast--End of Year 2012					
3		Cell Address	Recession- High Inflation	Recession- Low Inflation	Boom- High Inflation	Boom- Low Inflation
5	Changing Cells:					
6	Economic Outlook: R-Recession, B-Boom	$C$8	R	R	B	B
7	Inflation: H-High, L-Low	$C$9	H	L	H	L
8	Result Cells:					
9	Net Income after Taxes (Expansion)	$C$12	$69,974	$78,510	$61,316	$68,867
10	End-of-year Cash on Hand (Expansion)	$C$13	$84,974	$93,510	$76,316	$83,867
11	Net Income after Taxes (No Expansion)	$C$14	$72,601	$80,480	$63,526	$70,420
12	End-of-year Cash on Hand (No Expansion)	$C$15	$87,601	$95,480	$78,526	$85,420
13	Notes: Current Values column represents values of changing cells at					
14	time Scenario Summary Report was created. Changing cells for each					
15	scenario are highlighted in gray.					

FIGURE C-29   Scenario Summary for End of Year 2012

As you can see from these results, *not* expanding the business yields slightly better financial results at the end of year 2012. As the original Scenario Summary points out, it will take two years for the business expansion to start making more money when compared with not expanding. You can also revise the original spreadsheet to copy the columns out to 2014, 2015, and beyond to forecast future income and cash flows. However, note that the accuracy of a forecast gets worse as you extend it in time.

Managers must also maintain a healthy skepticism about the validity of their assumptions when formulating a DSS model. Most assumptions about economic outlooks, inflation, and interest rates are really educated guesses. For example, who could have predicted the economic meltdown in 2007? Business DSS models for investments, new product launches, business expansion, or projects commonly look at three possible outcomes: best case, most likely, and worst case. The most likely outcome is based on previous years' data already collected by the firm. The best-case and worst-case outcomes are formulated based on some percentage of performance that falls above or below the most likely scenario. At least these are data-driven forecasts, or what people in the business world call "Guessing—with data."

So, how do you reduce risk when making financial decisions based on DSS model results? It helps to formulate the model based on valid data, and to use conservative estimates for success. More importantly, collecting pertinent data and tracking the business results *after* deciding to invest or expand can help reduce the risk of failure for the enterprise.

Summary Sheets

When you start working on the Scenario Manager spreadsheet cases later in this book, you will need to know how to manipulate summary sheets and their data. Some of these operations are explained in the following sections.

Rerunning Scenario Manager

The Scenario Summary sheet does not update itself when you change formulas or inputs in the spreadsheet. To get an updated Scenario Summary, you must rerun Scenario Manager, as you did when changing the outputs from 2013 to 2012. Click the Summary button in the Scenario Manager dialog box, and then click OK. Another summary sheet is created; Excel numbers them sequentially (Scenario Summary, Scenario Summary 2, etc.), so you do not have to worry about Excel overwriting any of your older summaries. That is why you should rename each summary with a description of the changes.

Deleting Unwanted Scenario Manager Summary Sheets

When working with Scenario Manager, you might produce summary sheets you do not want. To delete an unwanted sheet, move your mouse to the group of sheet tabs at the bottom of the screen and *right*-click the tab of the sheet you want to delete. Click Delete from the menu that appears (see Figure C-30).

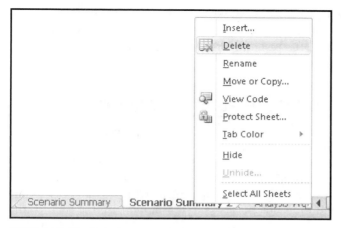

FIGURE C-30   Deleting unwanted worksheets

Charting Summary Sheet Data

You can easily chart Summary Sheet results using the Charts group in the Insert tab, as discussed in Tutorial E. Figure C-31 shows a clustered column chart prepared from the data in the Scenario Summary for 2013. Charts are useful for showing a visual comparison of financial results depending on economic conditions. As the chart shows, the best economic climate for the thrift shop is a Recession with Low Inflation.

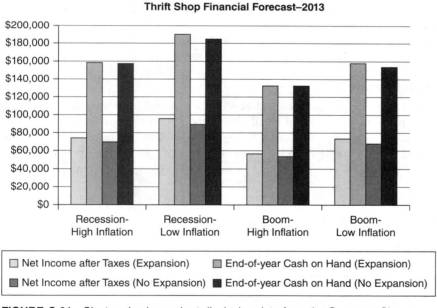

FIGURE C-31   Clustered column chart displaying data from the Summary Sheet

Copying Summary Sheet Data to the Clipboard

As you can with almost everything else in Microsoft Office, you can copy summary sheet data to other Office applications (a Word document or PowerPoint slide, for example) by using the Clipboard. Follow these steps:

1. Select the data range you want to copy.
2. Right-click the mouse and select Copy from the resulting menu.

3. Open the Word document or PowerPoint presentation into which you want to copy.
4. Click your cursor where you want the upper-left corner of the copied data to be displayed.
5. Right-click the mouse and select Paste from the resulting menu. The data should now appear on your document.

PRACTICE EXERCISE—TED AND ALICE'S HOUSE PURCHASE DECISION

Ted and Alice are a young couple who have been living in an apartment for the first two years of their marriage. They would like to buy their first house, but do not know if they would be able to make ends meet. Ted works as a carpenter's apprentice, and Alice is a customer service specialist at a local bank. In 2011, Ted's "take home" wages (after taxes and deductions) were $24,000, and Alice's take-home salary was $30,000. Ted gets a 2% raise every year, and Alice gets a 3% raise. Their apartment rent is $1,200 per month ($14,400 per year), but the lease is up for renewal and the landlord has said he will increase the rent for the next lease.

Ted and Alice have been looking at houses and have found one that they can buy, but they will need to borrow $200,000 for a mortgage. Their parents are helping them with the down payment and closing costs. After talking to several lenders, Ted and Alice have learned that the state legislature is voting on a first-time home buyers mortgage bond. If the bill passes, they will be able to get a 30-year fixed mortgage at 3% interest. Otherwise, they will have to pay 6% interest on the mortgage.

Because of the depressed housing market, Ted and Alice are not figuring equity value into their calculations. In addition, although the mortgage interest and real estate taxes will be deductible on their income taxes, those deductions will not be higher than the standard allowable tax deduction, so they are not figuring on any savings there either. Ted and Alice's other living expenses (such as car payments, food, and medical bills), the utilities expenses for either renting or buying, and estimated house maintenance expenses are listed in the Constants section (see Figure C-32).

Ted and Alice's primary concern is their cash on hand at the end of years 2012 and 2013. They are thinking of starting a family, but they know it will be difficult without adequate savings.

Getting Started on the Practice Exercise

If you closed Excel after the first tutorial exercise, start Excel again—it will automatically open a new workbook for you. If your Excel workbook from the first tutorial is still open, you may find it useful to start a new worksheet in the same workbook. Then you can refer back to the first tutorial when you need to structure or format the spreadsheet; the formatting of both exercises in this tutorial is similar. Set up your new worksheet as explained in the following sections.

Constants Section

Your spreadsheet should have the constants shown in Figure C-32. An explanation of the line items follows the figure.

	A	B	C	D
1	**Tutorial Exercise Skeleton--Ted and Alice's House Decision**			
2				
3	**Constants**	**2011**	**2012**	**2013**
4	Non-Housing Living Expenses (Cars, Food, Medical, etc)	NA	$36,000	$39,000
5	Mortgage Amount for Home Purchase	NA	$200,000	NA
6	Real Estate Taxes and Insurance on Home	NA	$3,000	$3,150
7	Utilities Expense (Heat & Electric)--Apartment	NA	$2,000	$2,200
8	Utilities Expense(Heat, Electric, Water, Trash)--House	NA	$2,500	$2,600
9	House Repair and Maintenance Expenses	NA	$1,200	$1,400

FIGURE C-32   Constants section

- Non-Housing Living Expenses—This value represents Ted and Alice's estimate of all their other living expenses for 2012 and 2013.
- Mortgage Amount for Home Purchase
- Real Estate Taxes and Insurance on Home—A lender has given Ted and Alice estimates for these values; they are usually paid monthly with the house mortgage payment. The money is placed in an escrow account and then paid by the mortgage company to the state or county and insurance company.
- Utilities Expense—Apartment—This value is Ted and Alice's estimate for 2012 and 2013 based on their 2011 bills.
- Utilities Expense—House—Currently the apartment rent includes fees for water, sewer, and trash disposal. If they get a house, Ted and Alice expect the utilities to be higher.
- House Repair and Maintenance Expenses—In an apartment, the landlord is responsible for repair and maintenance. Ted and Alice will have to budget for repair and maintenance on the house.

Inputs Section

Your spreadsheet should have the inputs shown in Figure C-33. An explanation of line items follows the figure.

	A	B	C	D
11	**Inputs**	**2011**	**2012**	**2013**
12	Rental Occupancy (H=High, L=Low)	NA		NA
13	First Time Buyer Bond Loans Available (Y=Yes, N=No)	NA		NA

FIGURE C-33 Inputs section

- Rental Occupancy (H=High, L=Low)—When the housing market is depressed (in other words, people are not buying homes), rental housing occupancy percentages are high, which allows landlords to charge higher rents when leases are renewed. Ted and Alice think their rent will increase in 2012. The amount of the increase depends on the Rental Occupancy. If the occupancy is high, Ted and Alice expect to see a 10% increase in rent in both 2012 and 2013. If occupancy is low, they only expect a 3% increase.

- First Time Buyer Bond Loans Available (Y=Yes, N=No)—As described earlier, when housing markets are depressed, local governments will frequently pass a bond bill to provide low-interest mortgage money to first-time home buyers. If the bond loans are available, Ted and Alice can obtain a 30-year fixed mortgage at only 3%, which is half the interest rate they would otherwise pay for a conventional mortgage.

Summary of Key Results Section

Figure C-34 shows what key results Ted and Alice are looking for. They want to know their End-of-year Cash on Hand for both 2012 and 2013 if they decide to stay in the apartment and if they decide to purchase the house.

	A	B	C	D
15	**Summary of Key Results**	**2011**	**2012**	**2013**
16	End-of-year Cash on Hand (Rent)	NA		
17	End-of-year Cash on Hand (Buy)	NA		

FIGURE C-34 Summary of Key Results section

These results are copied from the End-of-year Cash on Hand sections of the Income and Cash Flow Statements sections (both for renting and buying).

Calculations Section

Your spreadsheet will need formulas to calculate the apartment rent, house payments, and the interest rate for the mortgage (see Figure C-35). You will use the rent and house payments later in the Income and Cash Flow Statements for both renting and buying.

	A	B	C	D
19	**Calculations**	**2011**	**2012**	**2013**
20	Apartment Rent	$14,400		
21	House Payments	NA		
22	Interest Rate for House Mortgage		NA	NA

FIGURE C-35   Calculations section

- Apartment Rent—The 2011 amount is given. Use IF formulas to increase the rent by 10% if occupancy rates are high, or by 3% if occupancy rates are low.
- House Payments—This value is the total of the 12 monthly payments made on the mortgage. An important point to note is that house mortgage interest is always compounded *monthly*, not annually as in the thrift shop tutorial. To properly calculate the house payments for the year, you divide the annual interest rate by 12 to determine the monthly interest. You also have to multiply a 30-year mortgage by 12 to get 360 payments, and then multiply the PMT formula by 12 to get the total amount for your annual house payments. Also, you will precede the PMT function with a negative sign to make the payment amount a positive number. Your formula should look like the following:
=-PMT(B22/12,360,C5)*12
- Interest Rate for House Mortgage—Use the IF formula to enter a 3% interest rate if the bond money is available, and a 6% interest rate if no bond money is available.

Income and Cash Flow Statements Sections

As with the thrift shop tutorial, you want to see the Income and Cash Flow Statements for two scenarios—in this case, for continuing to rent and for purchasing a house. Each section ends with cash on hand at the end of 2011. As you can see in Figure C-36, Ted and Alice only have $4,000 in their savings.

	A	B	C	D
24	**Income and Cash Flow Statement (Continue to Rent)**	**2011**	**2012**	**2013**
25	Beginning-of-year Cash on Hand	NA		
26	Ted's Take Home Wages	$24,000		
27	Alice's Take Home Salary	$30,000		
28	Total Take Home Income	$54,000		
29	*Apartment Rent*	NA		
30	Utilities (Apartment)	NA		
31	Non-Housing Living Expenses	NA		
32	Total Expenses	NA		
33	End-of-year Cash on Hand	$4,000		
34				
35	**Income and Cash Flow Statement (Purchase House)**	**2011**	**2012**	**2013**
36	Beginning-of-year Cash on Hand	NA		
37	Ted's Take Home Wages	$24,000		
38	Alice's Take Home Salary	$30,000		
39	Total Take Home Income	$54,000		
40	*House Payments*	NA		
41	*Real Estate Taxes and Insurance*	NA		
42	Utilities (House)	NA		
43	*House Repair and Maintenance Expense*	NA		
44	Non-Housing Living Expenses	NA		
45	Total Expenses	NA		
46	End-of-year Cash on Hand	$4,000		

FIGURE C-36   Income and Cash Flow Statements sections (for both rent and purchase)

- Beginning-of-year Cash on Hand—This value is the End-of-year Cash on Hand from the previous year.
- Ted's Take Home Wages—This value is given for 2011. To get values for 2012 and 2013, increase Ted's wages by 2% each year.
- Alice's Take Home Salary—This value is given for 2011. To get values for 2012 and 2013, increase Alice's salary by 3% each year.
- Total Take Home Income—The sum of Ted and Alice's pay.
- Apartment Rent—The rent is copied from the Calculations section.
- House Payments—The house payments are also copied from the Calculations section.
- Real Estate Taxes and Insurance, Utilities (Apartment or House), House Repair and Maintenance Expense, and Non-Housing Living Expenses—These values all are copied from the Constants section.
- Total Expenses—This value is the sum of all the expenses listed above. Note that the house payment is now a positive number, so you can sum it normally with the other expenses.
- End-of-year Cash on Hand—This value is the Beginning-of-year Cash on Hand plus the Total Take Home Income minus the Total Expenses.

Scenario Manager Analysis

When you have completed the spreadsheet, set up Scenario Manager and create a Scenario Summary sheet. Ted and Alice want to look at their End-of-year Cash on Hand in 2013 for renting or buying under the following four scenarios:

- High occupancy and bond money available
- High occupancy and no bond money available
- Low occupancy and bond money available
- Low occupancy and no bond money available

If you have done your spreadsheet and Scenario Manager correctly, you should get the results shown in Figure C-37.

	A	B	C	D	E	F
1	Scenario Summary--Ted & Alice's House Purchase Decision					
2			Hi Occ- Bond $	Hi Occ- No Bond $	Lo Occ- Bond $	Lo Occ- No Bond $
4	**Changing Cells:**					
5	**Rental Occupancy (H-High, L-Low)**	$C$12	H	H	L	L
6	**Bond Mortgage Available (Y or N)**	$C$13	Y	N	Y	N
7	**Result Cells:**					
8	**End of Year Cash on Hand (Rent)**	$D$16	$4,505	$4,505	$7,016	$7,016
9	**End of Year Cash on Hand (Buy)**	$D$17	$7,090	($1,452)	$7,090	($1,452)
10	Notes: Current Values column represents values of changing cells at					
11	time Scenario Summary Report was created. Changing cells for each					
12	scenario are highlighted in gray.					

FIGURE C-37   Scenario Summary results

Interpreting the Results

Based on the Scenario Summary results, what should Ted and Alice do? At first glance, it looks like the safe decision is to stay in the apartment. Actually, their decision hinges on whether they can get the lower-interest mortgage from the first-time buyers' bond issue. If they can, and if occupancy levels in apartments stay high, purchasing a house will give them about $2,500 more in savings at the end of 2013 than if they continued renting. Some other intangible factors are that home owners do not need permission to have pets, detached houses are quieter than apartments, and homes usually have a yard for pets and children to play in. Also, for the purposes of this exercise, you did not consider the tax benefits of home ownership. Depending on the amount of mortgage interest and real estate taxes Ted and Alice have to pay, they may be able to itemize their deductions and pay less income tax. If the income tax savings are more than $1,500, they can purchase the house even at the higher interest rate. In any case, because you did the DSS model for them, Ted and Alice now have a quantitative basis to help them make a good decision.

Visual Impact: Charting the Results

Charts and graphs often add visual impact to a Scenario Summary. Using the data from the Scenario Summary output table, try to create a chart similar to the one in Figure C-38 to illustrate the financial impact of each outcome.

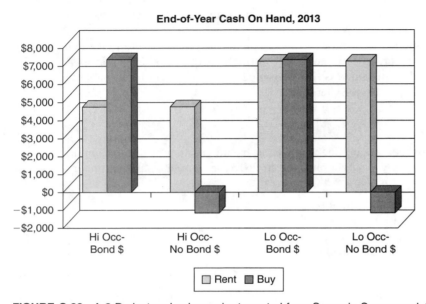

FIGURE C-38   A 3-D clustered column chart created from Scenario Summary data

Printing and Submitting Your Work

Ask your instructor which worksheets need to be printed for submission. Make sure your printouts of the spreadsheet, the Scenario Manager table, and the graph (if you created one) fit on one printed page apiece. Click the File tab on the Ribbon, click the Print icon, and then click Page Setup at the bottom of the Print Navigation pane. When the Page Setup dialog box opens, click the Page tab if it is not already open, then click the Fit to radio button and select the setup that is 1 page wide by 1 page tall. Your spreadsheet, table, and graph will be fitted to print on one page apiece.

REVIEW OF EXCEL BASICS

This section reviews some basic operations in Excel and some tips for good work practices. Then you will work through some cash flow calculations. Working through this section will help you complete the spreadsheet cases in the following chapters.

Save Your Work Often—and in More Than One Place

To guard against data loss in case of power outages, computer crashes, and hard drive failure, it is always a good idea to save your work to a separate storage device. Copying a file into two separate folders on the same hard disk is *not* an adequate safeguard. If you are working on your college's computer network and you have assigned network storage, the network storage is usually "mirrored"; in other words, it has duplicate drives recording data to prevent data loss if the system goes down. However, most laptops and home computers lack this feature. An excellent way to protect your work from accidental deletion is to purchase a USB "thumb" drive and copy all of your files to it.

When you save your Excel files, Windows will usually store them in the Documents folder unless you specify the storage location. Instead of just clicking the Save icon, a good idea is to click the File group in the Ribbon (or the Office Button in Office 2007), then click the Save As icon. A dialog box will appear with icons on the left side, as shown in Figure C-39. If you have previously saved your file to a particular location, it will appear in the Save in text box at the top of the dialog box. To save the file in the same location, click Save. If your work is stored elsewhere, you can find the location using the icons on the left side of the dialog box. If you are saving to a USB thumb drive, it will appear as a storage device when you click the Computer icon. Click the folder where you want to save your file.

NOTE

If you are trying an operation that might damage your spreadsheet and you do not want to use the Undo command, you can use the Save As command, and then add a number or letter to the filename to save an additional copy to "play with." Your original work will be preserved.

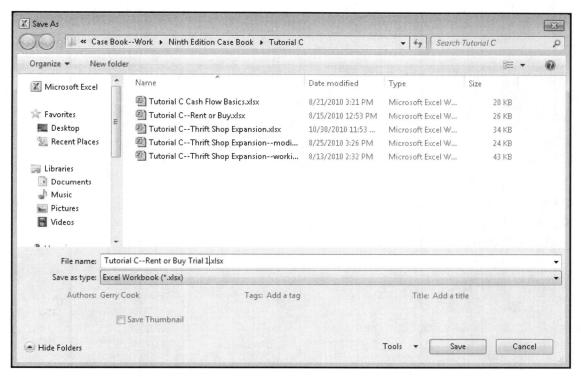

FIGURE C-39   The Save As dialog box

You will review the following topics: formatting cells, displaying the spreadsheet cell formulas, circular reference errors, using the AND and OR logical operators in IF statements, and using nested IF statements to produce more than two outcomes.

Formatting Cells

Cell Alignment

Headings for columns are usually centered, while numbers in cells are usually aligned to the right. To set the alignment of cell data:

1. Highlight the cell or cell range to format.
2. Select the Home tab.
3. In the Alignment group, click the button representing the horizontal alignment you want for the cell (Left Align, Center, or Right Align).
4. Also in the Alignment group, above the horizontal alignment buttons, click the vertical alignment you want (Top Align, Middle Align, or Bottom Align). Middle Align is the most common vertical alignment for cells.

Cell Borders

Bottom borders are common for headings, and accountants include borders and double borders to indicate subtotals and grand totals on spreadsheets. Sometimes it is also useful to put a "box" border around a table of values or a section of a spreadsheet. To create borders:

1. Highlight the cell or cell range that needs a border.
2. Select the Home tab.
3. In the Font group, click the drop-down arrow of the Border icon. A menu of border selections appears (see Figure C-40).
4. Choose the desired border for the cell or group of cells. Note that All Borders creates a box border around each cell, while Outside Borders draws a box around a group of cells.

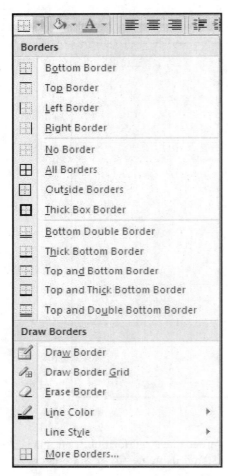

FIGURE C-40 Selections in the Borders menu

Number Formats

For financial numbers, you usually use the Currency format. (Do not use the Accounting format, as it places the $ sign to the far left side of the cell.) To apply the appropriate Currency format:

1. Highlight the cell or cell range to be formatted.
2. Select the Home tab.
3. In the Number group, select Currency in the Number Format drop-down list.
4. To set the desired number of decimal places, click the Increase Decimal or Decrease Decimal button in the bottom-right corner of the group (see Figure C-41).

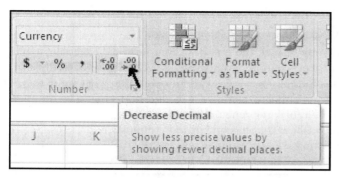

FIGURE C-41   Increase Decimal and Decrease Decimal buttons

If you do not know what a button does in Office, hover your mouse pointer over the button to see a description.

Format "Painting"

If you want to copy *all* the format properties of a certain cell to other cells, use the Format Painter. First, select the cell whose format you want to copy. Then click the Format Painter button (the paintbrush icon) in the Clipboard Group under the Home tab (see Figure C-42). When you click the button, the mouse pointer turns into a paintbrush. Click the cell you want to reformat to copy all the formatting to it. To format multiple cells, select the cell whose format you want to copy, and then click *twice* on the Format Painter button. The mouse cursor will become a paintbrush, and the paint function will stay on so you can reformat as many cells as you want. To turn off the Format Painter, click its button again or press the Esc key.

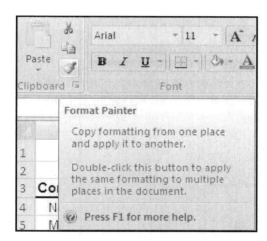

FIGURE C-42   The Format Painter button

Showing the Excel Formulas in the Cells

Sometimes your instructor might want you to display or print the formulas in the spreadsheet cells. If you want the spreadsheet cells to display the actual cell formulas, follow these steps:

1. While holding down the Ctrl key, press the key in the upper-left corner of the keyboard that contains the back quote (`) and tilde (~). The spreadsheet will display the formulas in the cells. The columns may also become quite wide—if so, do not resize them.
2. The Ctrl+`~ key combination is also a toggle. To restore your spreadsheet back to the normal cell contents, press Ctrl+`~ again.

Understanding Circular Reference Errors

When entering formulas, you might make the mistake of referring to the cell in which you are entering the formula as part of the formula, even though it should only display the output of that formula. Referring a cell back to itself in a formula is called a *circular reference*. For example, say that in cell B2 of a worksheet, you enter =B2-B1. Excel cannot calculate the answer to this formula because it cannot determine what value to enter in the formula for B2. A terrible but apt analogy for a circular reference is a cannibal trying to eat himself! Fortunately, Excel informs you when you try to enter a circular reference into a formula (see Figure C-43). Excel also warns you if you try to open an existing spreadsheet that has one or more circular references. Before you can use the spreadsheet, you must fix the formulas that contain circular references.

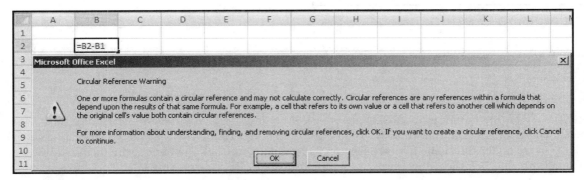

FIGURE C-43   Excel circular reference warning

Using AND and OR Functions in IF Statements

Recall that the IF function has the following syntax:

IF(test condition, result if test is True, result if test is False)

The test conditions in the previous example IF statements tested only one cell's value, but a test condition can test more than one value of a cell.

For example, look at the thrift shop tutorial again. The Total Sales Dollars for 2012 depended on the economic outlook (recession or boom). The original IF statement was =IF($C$8="R",B18*1.3,B18*1.15), as shown in Figure C-44. This function increased the 2011 Total Sales Dollars by 30% if there was continued recession, but only increased the total by 15% if there was a boom.

C18		f_x =IF($C$8="R",B18*1.3,B18*1.15)		
	A	B	C	D
7	**Inputs**	**2011**	**2012**	**2013**
8	Economic Outlook (R=Recession, B=Boom)	NA	R	NA
9	Inflation Outlook (H=High, L=Low)	NA	H	NA
10				
11	**Summary of Key Results**	**2011**	**2012**	**2013**
12	Net Income after Taxes (Expansion)	NA	$69,974	$73,660
13	End-of-year Cash on Hand (Expansion)	NA	$84,974	$158,634
14	Net Income after Taxes (No Expansion)	NA	$72,601	$69,936
15	End-of-year Cash on Hand (No Expansion)	NA	$87,601	$157,537
16				
17	**Calculations (Expansion)**	**2011**	2012	**2013**
18	Total Sales Dollars	$350,000	$455,000	$591,500
19	Cost of Goods Sold	$245,000	$337,610	$465,227
20	Cost of Goods Sold (as a percent of Sales)	70%	74%	79%
21	Interest Rate for Business Loan	5%	NA	NA

FIGURE C-44   The original IF statement used to calculate the Total Sales Dollars for 2012

To take the IF argument one step further, assume that the Total Sales Dollars for 2012 depended not only on the Economic Outlook, but on the Inflation Outlook (High or Low). Say there are two possibilities:

- Possibility 1: If the economic outlook is for a Recession and the inflation outlook is High, the Total Sales Dollars for 2012 will be 30% higher than in 2011.
- Possibility 2: For the other three cases (Recession and Low Inflation, Boom and High Inflation, and Boom and Low Inflation), assume that the Total Sales Dollars will only be 15% higher than in 2011.

The first possibility requires two conditions to be true at the same time: C8="R" and C9="H". You can include an AND( ) function inside the IF statement to reflect the additional condition as follows:

=IF(AND($C$8="R",$C$9="H"), B18*130%,B18*115%)

When the test argument uses the AND() function, conditions "R" *and* "H" both must be present at the same time for the statement to use the true result (multiplying last year's sales by 130%). Any of the other three outcome combinations will cause the statement to use the false result (multiplying last year's sales by 115%).

You can also use an OR() function in an IF statement. For example, assume that instead of both conditions (Recession and High Inflation) having to be present, only one of the two conditions needs to be present for sales to increase by 30%. In this case, you use the OR() function in the test argument as follows:

=IF(OR($C$8="R",$C$9="H"), B18*130%,B18*115%)

In this case, if *either* of the two conditions (C8="R" or C9="H") is true, the function will return the true argument, multiplying the 2011 sales by 130%. If *neither* of the two conditions is true, then the function will return the false argument, multiplying the 2011 sales by 115% instead.

Using IF Statements Inside IF Statements (Also Called "Nesting Ifs")

By now you should be familiar with IF statements, but here is a quick review of the syntax:

=IF(test condition, result if test is True, result if test is False)

In the preceding examples, only two courses of action were possible for each of the inputs: Recession or Boom, High Inflation or Low Inflation, Rental Occupancy High or Low, Bond Money Available or No Bond Money Available. The tutorial used only two possible outcomes to keep them simple.

However, in the business world, decision support models are frequently based on three or more possible outcomes. For capital projects and new product launches, you will frequently project financial outcomes based on three possible scenarios: Most Likely, Worst Case, and Best Case. You can modify the IF statement by placing another IF statement inside the result argument if the first test is false, creating the ability to launch two more alternatives from the second IF statement. This is called "nesting" your IF statements.

Try a simple nested IF statement first: In your thrift shop example, assume that three economic outlooks are possible: Recession (R), Boom (B), or Stable (S). As before, the 2012 Total Sales Dollars (cell C18) will be the 2011 Total Sales Dollars increased by some fixed percentage. In a Recession, sales will increase by 30%, in a Boom they will increase by 15%, and for a Stable Economic Outlook, sales will increase by 22%, which is roughly midway between the other two percentages. You can "nest" the IF statement in cell C18 to reflect the third outcome as follows:

=IF($C$8= "R",B18*130%,IF($C$8="B",B18*115%,B18*122%))

Note the added IF statement inside the False value argument. You can break down this statement:

- If the value in cell C8 is "R", multiply the value in cell B18 by 130%, and enter the result in cell C18.
- If the value in cell C8 is not "R", check whether the value in cell C8 is "B". If it is, multiply the value in cell B18 by 115%, and enter the result in cell C18.
- If the value in cell C8 is not "B", multiply the value in cell B18 by 122%, and enter the result in cell C18.

If you have four or more alternatives, you can keep nesting IF statements inside the false argument for the outer IF statements. (Excel 2007 and later versions have a limit of 64 levels of nesting in the IF function, which should take care of every conceivable situation.)

NOTE

The "embedded IFs" in a nested IF statement are not preceded by an equals sign. Only the first IF gets the equals sign.

Cash Flow Calculations: Borrowing and Repayments

The Scenario Manager cases that follow in this book require accounting for money that the fictional company will have to borrow or repay. This money is not like the long-term loan that the Collegetown Thrift Shop is considering for its expansion. Instead, this money is short-term borrowing that companies use to pay current obligations, such as purchasing inventory or raw materials. Such short-term borrowing is called a line of credit, and is extended to businesses by banks, much like consumers have credit cards. Lines of credit usually involve interest payments, but for simplicity's sake, focus instead on how to do short-term borrowing and repayment calculations.

To work through cash flow calculations, you must make two assumptions about a company's borrowing and repayment of short-term debt. First, you assume that the company has a desired *minimum* cash level at the end of a fiscal year (which is also its cash level at the start of the next fiscal year), to ensure that the company can cover short-term expenses and purchases. Second, assume the bank that serves the company will provide short-term loans (a line of credit) to make up the shortfall if the end-of-year cash falls below the desired minimum level.

NCP stands for Net Cash Position, which equals beginning-of-year cash plus net income after taxes for the year. NCP represents the available cash at the end of the year, *before* any borrowing or repayment. For the three examples shown in Figure C-45, set up a simple spreadsheet in Excel and determine how much the company needs to borrow to reach its minimum year-end cash level. Use the IF function to enter 0 under Amount to Borrow if the company does not need to borrow any money.

	A	B	C	D
1	Example	NCP	Minimum Cash Required	Amount to Borrow
2	1	$25,000	$10,000	?
3	2	$9,000	$10,000	?
4	3	($12,000)	$10,000	?

FIGURE C-45   Examples of borrowing

You can also assume that the company will use some of its cash on hand at the end of the year to pay off as much of its outstanding debt as possible without going below its minimum cash on hand required. The "excess" cash is the company's NCP *less* the minimum cash on hand required—any cash above the minimum is available to repay any debt. In the examples shown in Figure C-46, compute the excess cash and then compute the amount to repay. In addition, compute the ending cash on hand after the debt repayment.

	A	B	C	D	E	F
9	Example	NCP	Minimum Cash Required	Beginning-of-Year Debt	Repay?	Ending Cash
10	1	$12,000	$10,000	$5,000	?	?
11	2	$13,000	$10,000	$8,000	?	?
12	3	$20,000	$10,000	$0	?	?
13	4	$60,000	$10,000	$40,000	?	?
14	5	($20,000)	$10,000	$10,000	?	?

FIGURE C-46   Examples of debt repayment

In the Scenario Manager cases of the following chapters, your spreadsheet will need two bank financing sections beneath the Income and Cash Flow Statements sections. You will build the first section to calculate any needed borrowing or repayment at year's end to compute year-end cash on hand. The second section will calculate the amount of debt owed at the end of the year after any borrowing or debt repayment.

Return to the Collegetown Thrift Shop tutorial and assume that it includes a line of credit at a local bank for short-term cash management. The first new section extends the end-of-year cash calculation, which was shown for the thrift shop in Figure C-19. Figure C-47 shows the structure of the new section highlighted in boldface.

	A	B	C	D
28	**Income and Cash Flow Statements (Expansion)**	**2011**	**2012**	**2013**
29	Beginning-of-year Cash on Hand	NA	$15,000	$0
30	Sales (Revenue)	NA	$455,000	$591,500
31	Cost of Goods Sold	NA	$337,610	$465,227
32	*Business Loan Payment*	NA	($12,950)	($12,950)
33	Income before Taxes	NA	$104,440	$113,323
34	Income Tax Expense	NA	$34,465	$39,663
35	Net Income after Taxes	NA	$69,974	$73,660
36	**Net Cash Position NCP** **Beginning-of-year Cash on Hand** **plus Net Income after Taxes**	NA	$84,974	$73,660
37	**Line of credit borrowing from bank**	NA		
38	**Line of credit repayments to bank**	NA		
39	**End-of Year Cash on Hand**	**$15,000**		

FIGURE C-47   Calculation section for End-of-Year Cash on Hand with borrowing and repayments added

The heading in cell A36 was originally End-of-year Cash on Hand in Figure C-19, but you will add line-of-credit borrowing and repayment to the end-of-year totals. You must add the line-of-credit borrowing from the bank to the NCP and subtract the line-of-credit repayments to the bank from the NCP to obtain the End-of-Year Cash on Hand.

The second new section you add will compute the End-of year debt owed. This section is called Debt Owed, as shown in Figure C-48.

	A	B	C	D
42	**Debt Owed**	**2011**	**2012**	**2013**
43	**Beginning-of-year debt owed**	NA		
44	**Borrowing from bank line-of-credit**	NA		
45	**Repayment to bank line-of-credit**	NA		
46	**End-of-year debt owed**	**$47,000**		

FIGURE C-48   Debt Owed section

As you can see, the thrift shop currently owes $47,000 on its line of credit at the end of 2011. The End-of-year debt owed equals the Beginning-of-year debt owed plus any new borrowing from the bank's line of credit, minus any repayment to the bank's line of credit. Therefore, the formula in cell C46 would be:

=C43+C44-C45

Assume that the amounts for borrowing and repayment (cells C44 and C45) were calculated in the first new section (for the year 2012, the amounts would be in cells C38 and C39), and then copied into the second section. The formula for cell C44 would be =C38, and for cell C45 would be =C39. The formula for cell C43, Beginning-of-year debt owed in 2012, would simply be the End-of-year debt owed in 2011, or =B46.

Now that you have added the spreadsheet entries for borrowing and repayment, consider the logic for the borrowing and repayment formulas.

Calculation of Borrowing from the Bank Line of Credit

When using logical statements, it is sometimes easier to state the logic in plain language and then turn it into an Excel formula. For borrowing, the logic in plain language is:

If (cash on hand before financing transactions is greater than the minimum cash required,

then borrowing is not needed; else,

borrow enough to get to the minimum)

You can restate this logic as the following:

If (NCP is greater than minimum cash required,

then borrowing from bank=0;

else, borrow enough to get to the minimum)

You have not added minimum cash at the end of the year as a requirement, but you could add it to the Constants section at the top of the spreadsheet (in this case the new entry would be cell C6). Assume that you want $50,000 as the minimum cash on hand at the end of both 2012 and 2013. Assuming that the NCP is shown in cell C37, you could restate the formula for borrowing (cell C38) as the following:

IF(NCP>Minimum Cash, 0; otherwise, borrow enough to get to the minimum cash)

You have cell addresses for NCP (cell C37) and for Minimum Cash (cell $C$6). To develop the formula for cell C38, substitute the cell address for the test argument; the true argument is simply zero (0), and the false argument is the minimum cash minus the current NCP. The formula stated in Excel for cell C38 would be:

=IF(C37>=$C$6, 0, $C$6-C37)

Calculation of Repayment to the Bank Line of Credit

Simplify the statements first in plain language:

IF(beginning of year debt=0, repay 0 because nothing is owed), but

IF(NCP is less than the minimum, repay 0, because you must borrow), but

IF(extra cash equals or exceeds the debt, repay the whole debt),

ELSE (to stay above the minimum cash, repay the extra cash above the minimum)

Look at the following formula. If you assume that the repayment amount will be in cell C39, the beginning-of-year debt is in cell C43, and the minimum cash target is still in cell $C$6, the repayment formula for cell C39 with the nested IFs should look like the following:

=IF(C43=0,0,IF(C37<=$C$6,0,IF(C37-$C$6>=C43,C43,C37-$C$6)))

The new sections of the thrift shop spreadsheet would look like those in Figure C-49.

C39		fx	=IF(C43=0,0,IF(C37<=$C$6,0,IF(C37-$C$6>=C43,C43,C37-$C$6)))	
	A	B	C	D
29	**Income and Cash Flow Statements (Expansion)**	**2011**	**2012**	**2013**
30	Beginning-of-year Cash on Hand	NA	$15,000	$50,000
31	Sales (Revenue)	NA	$455,000	$591,500
32	Cost of Goods Sold	NA	$337,610	$465,227
33	*Business Loan Payment*	NA	($12,950)	($12,950)
34	Income before Taxes	NA	$104,440	$113,323
35	Income Tax Expense	NA	$34,465	$39,663
36	Net Income after Taxes	NA	$69,974	$73,660
37	**Net Cash Position NCP** **Beginning-of-year Cash on Hand** **plus Net Income after Taxes**	NA	$84,974	$123,660
38	Line of credit borrowing from bank	NA	$0	$0
39	Line of credit repayments to bank	NA	$34,974	$12,026
40	**End-of Year Cash on Hand**	$15,000	$50,000	$111,634
41				
42	**Debt Owed**	**2011**	**2012**	**2013**
43	**Beginning-of-year debt owed**	NA	$47,000	$12,026
44	**Borrowing from bank line-of-credit**	NA	$0	$0
45	**Repayment to bank line-of-credit**	NA	$34,974	$12,026
46	**End-of-year debt owed**	$47,000	$12,026	$0

FIGURE C-49   Thrift shop spreadsheet with line-of-credit borrowing, repayments, and Debt Owed added

Notice that all the owed line-of-credit debt will be paid off by the end of 2013.

Answers to the Questions about Borrowing and Repayment

Figures C-50 and C-51 display solutions for the borrowing and repayment calculations.

	A	B	C	D
1	Example	NCP	Minimum Cash Required	Amount to Borrow
2	1	$25,000	$10,000	$0
3	2	$9,000	$10,000	$1,000
4	3	($12,000)	$10,000	$22,000

FIGURE C-50 Answers to examples of borrowing

In Figure C-50, the formula in cell D2 for the amount to borrow is =IF(B2>=C2,0,C2-B2).

	A	B	C	D	E	F
9	Example	NCP	Minimum Cash Required	Beginning-of-Year Debt	Repay?	Ending Cash
10	1	$12,000	$10,000	$5,000	$2,000	$10,000
11	2	$13,000	$10,000	$8,000	$3,000	$10,000
12	3	$20,000	$10,000	$0	$0	$20,000
13	4	$60,000	$10,000	$40,000	$40,000	$20,000
14	5	($20,000)	$10,000	$10,000	$0	NA

FIGURE C-51 Answers to examples of repayment

In Figure C-51, the formula in cell E10 for the amount to repay is

=IF(B10>=C10,IF(D10>0,MIN(B10-C10,D10),0),0).

Note the following points about the repayment calculations shown in Figure C-51.

- In Example 1, only $2,000 is available for debt repayment ($12,000 – $10,000) to avoid dropping below the Minimum Cash Required.
- In Example 2, only $3,000 is available for debt repayment.
- In Example 3, the Beginning-of-Year Debt was zero, so the Ending Cash is the same as the Net Cash Position.
- In Example 4, there was enough cash to repay the entire $40,000 debt, leaving $20,000 in Ending Cash.
- In Example 5, the company has cash problems—it cannot repay any of the Beginning-of-Year Debt of $10,000, and it will have to borrow an additional $30,000 to reach the Minimum Cash Required target of $10,000.

You should now have all the basic tools you need to tackle Scenario Manager in Cases 6 and 7. Good luck!

CASE **6**

FUTURE CARS INC. PRODUCT STRATEGY DECISION

Decision Support Using Excel

PREVIEW

Future Cars Inc. is a high-tech, alternative-energy automobile company based in the Northeast United States. The company has already designed and built an Ultra-Low Emission Vehicle (ULEV) that is selling well, but Future Cars must make a major decision about what kind of car to build and market in the future. The two basic car designs under consideration are an electric-gas hybrid vehicle similar to those on the market today, and a hydrogen fuel cell-powered vehicle that does not use gasoline.

Each design has its advantages and drawbacks. The hybrid technology is mature and proven, but it has the added complexity of a gasoline engine, a battery pack, and an electric motor in the drive power train. The fuel-cell technology is simpler in theory and requires fewer parts (the power train includes a fuel-cell stack, a battery pack, and an electric motor), but the technology is still experimental, and the parts are more expensive. The labor costs to assemble the fuel-cell vehicle will be lower than those of the electric hybrid.

From a marketing standpoint, the electric hybrid is not as innovative as the fuel-cell car, making it a "me too" product in a market with dozens of other hybrids. The fuel-cell car has the potential to generate more demand, particularly from environmentally conscious consumers. An ever-present influence on the demand for alternative-energy cars is the price of gasoline.

The Finance, Sales, and Operations managers of Future Cars Inc. have created some initial inputs for you to develop an investment decision model using an Excel workbook. Your finished decision model will be instrumental in helping Future Cars Inc. choose a wise product strategy.

PREPARATION

- Review the spreadsheet concepts discussed in class and in your textbook.
- Your instructor may assign Excel exercises to help prepare you for this case.
- Tutorial C has an excellent review of IF and nested IF statements that will help you with this case.
- Review the file-saving instructions—it is always a good idea to save an extra copy of your work on a USB thumb drive.
- Reviewing Tutorial F will help you brush up on your presentation skills.
- Because Future Cars Inc. uses a strategic investment decision model, you will calculate the internal rate of return (IRR) in the decision model. If you are unfamiliar with the IRR function in Excel, this case includes a section that explains how to set it up.

BACKGROUND

You are an information analyst for Future Cars Inc. The company president has asked you to prepare a quantitative analysis of financial, sales, and operations data to help determine which technology path would offer the best strategic opportunity for the firm. The department managers have been asked to provide the following data from their functional areas:

- Financial and Accounting—The current cash position of the company, the cash outlay for the two investment choices, data for manufacturing costs and period costs for each alternative, and the corporate income tax rates
- Sales—Forecasts for U.S. sales of alternative-energy cars, a formula for calculating the sales demand for each technology alternative, the effect of gasoline prices on the sales demand, and the projected market pricing for alternative-energy cars
- Operations—Labor hours required for manufacturing cars of either technology

Specifically, the departments have given you the following data:

- Direct labor costs per hour, 2013 through 2015
- Direct materials costs per hybrid or fuel-cell vehicle, 2013 through 2015
- Direct labor hours per hybrid or fuel-cell vehicle, 2013 through 2015
- Manufacturing overhead allocation rates per direct labor hour, 2013 through 2015
- Capital investment for electric hybrid and fuel-cell cars (end of 2012)
- Median alternative-energy car sales price for 2012 (market price)
- Period costs (2012)
- End-of-year cash on hand (2012)—This money is available for investment.

In addition, the Assignment 1 section contains information you will need to write the formulas for the Calculations section, Income and Cash Flow Statements section, and IRR Calculation section.

You will use Excel to see how much profit and positive cash flow the two alternatives will generate for Future Cars for the next three years, and you will use Excel to calculate an internal rate of return for each alternative. You will also examine the effects of the economy (recession or boom) and the price of gasoline (low or high) on your projected vehicle sales and profits for each alternative. In summary, your DSS will include the following inputs:

1. Your decision to invest in either the hybrid electric or fuel-cell technology
2. Whether the economic outlook is for a recession or boom cycle
3. Whether the price of gasoline is low or high

Because Future Cars is entering a market that already has substantial competition, the company cannot set the price for its automobiles simply by marking up the cost by a certain percentage. The company knows the market price for alternative-energy cars and must make its prices competitive. The new fuel-cell technology will allow Future Cars to price its fuel-cell cars slightly higher than the electric hybrid models, but the company knows it is still competing against well-established competitors with good products.

Your DSS model must account for the effects of the preceding three inputs on costs, selling prices, sales demand, and other variables. If you design it well, your model will let you develop "what-if" scenarios with all the inputs, see the results, and show a preferred alternative for Future Cars management to adopt.

ASSIGNMENT 1: CREATING A SPREADSHEET FOR DECISION SUPPORT

In this assignment, you will create a spreadsheet to model the business decision that Future Cars Inc. is seeking. In Assignment 2, you will write a report to the CEO about your analysis and recommendations. In Assignment 3, you will prepare and give a presentation of your analysis and recommendations.

You will start by creating the spreadsheet model of the company's financial and marketing data. The model will cover three years of sales (2013 through 2015) for the new technology selected. Assume that the preliminary research and development was completed in 2011 and 2012, and that engineering and manufacturing assets are in place to begin production and sales in 2013.

This section will help you set up each of the following spreadsheet components before entering the cell formulas:

- Constants
- Inputs
- Summary of Key Results

- Calculations
- Income and Cash Flow Statements
- Internal Rate of Return Calculation

Note that the Internal Rate of Return Calculation section is new to the spreadsheet model for this edition of *Problem Solving Cases*. The section was added because Excel financial formulas such as IRR work better if the cash outflow and inflow data is arranged in a vertical column with the years in ascending order, as opposed to taking the cash flows from across the page or from nonadjacent cells.

Constants Section

First, build the skeleton of your spreadsheet. Set up your Constants section, as shown in Figure 6-1. An explanation of the line items follows the figure.

	A	B	C	D	E
1	**Future Cars Inc. Strategic Investment Decision**				
2					
3	**Constants**	**2012**	**2013**	**2014**	**2015**
4	Direct Labor Cost/Hr, Future Cars Inc.	NA	$30	$32	$33
5	Direct Materials Cost per Electric Hybrid Car (3% inflation)	NA	$11,000	$11,330	$11,670
6	Direct Materials Cost per Fuel Cell Car (3% inflation)	NA	$14,000	$14,420	$14,853
7	Direct Labor Hours per Electric Hybrid Car (5% learning curve)	NA	160.00	152.00	144.40
8	Direct Labor Hours per Fuel Cell Car (5% learning curve)	NA	130.00	123.50	117.33
9	Overhead Allocation Rate per Direct Labor Hour (both technologies)	NA	$60.00	$61.80	$63.65
10	Capital Investment for Electric Hybrid Car	$600,000,000	NA	NA	NA
11	Capital Investment for Fuel Cell Car	$1,100,000,000	NA	NA	NA
12	Median Alternative Energy Car Sales Price, 2012	$35,000	NA	NA	NA
13	Corporate Income Tax Rate	25%	25%	26%	27%

FIGURE 6-1   Constants section

- Direct Labor Cost/Hr, Future Cars Inc.—This value is the cost per direct labor hour worked on the manufacturing line, to include payroll taxes and benefits.
- Direct Materials Cost per Electric Hybrid Car—This value is the cost of the parts needed to build one electric hybrid car, with a three percent inflation factor built in for each year.
- Direct Materials Cost per Fuel Cell Car—This value is the cost of the parts needed to build one fuel-cell car, with a three percent inflation factor built in for each year.
- Direct Labor Hours per Electric Hybrid Car—This value is the number of direct labor hours needed to build one electric hybrid car, with a five percent learning factor built in for each year. (The learning factor reflects the increasing efficiency of labor with experience.)
- Direct Labor Hours per Fuel Cell Car—This value is the number of direct labor hours needed to build one fuel-cell car, with a five percent learning factor built in for each year of production.
- Overhead Allocation Rate per Direct Labor Hour (both technologies)—This value is the estimated amount of manufacturing overhead costs, such as indirect materials, indirect labor, and depreciation of plant, maintenance, and repair materials. These costs are allocated to the cost of the car based on the number of direct labor hours needed to make the car. Overhead can be allocated in a number of ways, but Future Cars Inc. allocates on the basis of direct labor because the assembly process is labor-intensive.
- Capital Investment for Electric Hybrid Car—This value is the total amount of investment in retooled plant equipment and technology to produce the electric hybrid car.
- Capital Investment for Fuel Cell Car—This value is the total amount of investment in retooled plant equipment and technology to produce the fuel-cell car.
- Median Alternative Energy Car Sales Price, 2012—This value is the median selling price for current alternative-energy cars in 2012. This value is important for determining a competitive price for new Future Cars models.
- Corporate Income Tax Rate—These values are the predicted corporate income tax rates for Future Cars for the new technology. Note that these rates are lower than usual corporate income tax rates because Future Cars will receive R&D tax credits for its new technology.

Inputs Section

Your spreadsheet model must include the following inputs, which will apply for all three years, as shown in Figure 6-2.

	A	B	C	D	E
15	**Inputs**	**2012**	**2013**	**2014**	**2015**
16	Economic Outlook (R=Recession, B=Boom)	NA		NA	NA
17	Gasoline Price Outlook (L=Low, H=High)	NA		NA	NA
18	Technology Selection (E=Electric Hybrid, F=Fuel Cell)	NA		NA	NA

FIGURE 6-2   Inputs section

- Economic Outlook—This value is either Recession (R) or Boom (B). The economic outlook will affect car sales volume; people tend to buy less during a recession.
- Gasoline Price Outlook—This value is either Low (L) or High (H). The price of gasoline has a significant effect on what types of cars people buy. As gas prices rise, the demand for alternative-energy cars rises.
- Technology Selection—This value is the basic input for the strategic decision to build either the proven Electric Hybrid (E) technology or the more innovative Fuel Cell (F) technology.

Summary of Key Results Section

This section (see Figure 6-3) contains the results data, which is of primary interest to the management team at Future Cars Inc. It includes income and end-of-year cash on hand information, as well as the internal rate of return (annualized) for a particular set of business inputs. This section summarizes the values calculated from the Calculations, Income and Cash Flow Statements, and Internal Rate of Return Calculation sections.

	A	B	C	D	E
20	**Summary of Key Results**	**2012**	**2013**	**2014**	**2015**
21	Net Income After Taxes	NA			
22	End-of-year Cash on hand	NA			
23	Internal Rate of Return for Investment	NA	NA	NA	

FIGURE 6-3   Summary of Key Results section

For each year from 2013 to 2015, your spreadsheet should show net income after taxes (which will also be the cash inflows for the IRR calculation) and end-of-year cash on hand. Because Future Cars is funding the capital investment from its own cash on hand at the end of 2012, there is no debt to repay. However, the company wants to know the IRR at the end of 2015.

Calculations Section

The Calculations section includes the calculations you need to perform to determine the number of cars that Future Cars Inc. hopes to sell, the costs of making the cars, the period (nonmanufacturing) costs of running the new product segment, and how the company plans to price its cars for each year. See Figure 6-4.

	A	B	C	D	E
25	**Calculations**	**2012**	**2013**	**2014**	**2015**
26	New Product Capital Investment (internally financed)		NA	NA	NA
27	Total Forecasted U.S. Market Demand for Alternative Energy Cars	2,000,000			
28	Forecasted Sales for Future Cars (units)	NA			
29	Total Direct Materials	NA			
30	Total Direct Labor	NA			
31	Total Manufacturing Overhead	NA			
32	Cost of Goods Sold	NA			
33	Manufacturing Cost per Unit	NA			
34	Unit Sales Price	$35,000.00			
35	Period Costs (Sales, Marketing, Distribution, Customer Service)	$1,500,000,000			

FIGURE 6-4   Calculations section

- New Product Capital Investment—This value is the amount of investment money spent at the end of 2012, depending on the Technology Selection from the Inputs section. If the selected technology is Electric Hybrid (E), this cell should display the capital investment amount from the Constants section for the electric hybrid (the value in cell B10). If the selected technology is Fuel Cell (F), the cell should display the capital investment amount from the Constants section for the fuel-cell car (the value in cell B11). Note that when you see the word "if" in the text, you need to write a formula using the IF function for the target cell.

- Total Forecasted U.S. Market Demand for Alternative Energy Cars—You are given the forecasted U.S. market demand for alternative-energy cars for 2012. You will have to write formulas to calculate the forecasted U.S. market demand for 2013, 2014, and 2015. The Marketing Department thinks that the U.S. market demand for alternative-energy cars will depend on both the Economic Outlook and the Gasoline Price Outlook for the next three-year cycle. These outlooks have four possible combinations and results:

  - Recession and low gas prices—Market demand for each year will only be 95 percent of the previous year's demand.
  - Recession and high gas prices—Market demand for each year will only be 98 percent of the previous year's demand.
  - Boom and low gas prices—Market demand for each year will be exactly the same as the previous year's demand (or 100 percent of the previous year).
  - Boom and high gas prices—Market demand for each year will be two percent higher than the previous year's demand (or 102 percent of the previous year).

  (Hint: For these forecasts, you will have to use a nested IF statement that contains AND functions.)

- Forecasted Sales for Future Cars—The Marketing analysts have determined that Future Cars' market penetration for car sales will depend on the technology selected. If the hybrid electric technology is selected, Future Cars Inc. will capture nine percent of the U.S. market demand. If the fuel-cell technology is chosen, the company will capture 12 percent of the U.S. market demand because only one other car maker (Honda) currently offers a fuel-cell car.

- Total Direct Materials—This value is the total amount of money spent on direct materials to produce the forecasted number of cars that Future Cars will sell each year. If the electric hybrid is selected in the Inputs section, the Total Direct Materials is the direct materials for each electric hybrid car (from the Constants section) multiplied by the Forecasted Sales for Future Cars. If the fuel-cell technology is selected in the Inputs section, the Total Direct Materials is the direct materials for each fuel-cell car (from the Constants section) multiplied by the Forecasted Sales for Future Cars.

- Total Direct Labor—This value is the total amount of money spent on direct labor to produce the forecasted number of cars that Future Cars will sell each year. If the electric hybrid is selected in the Inputs section, the Total Direct Labor is the Direct Labor rate multiplied by the Direct Labor Hours for each electric hybrid car (both values are in the Constants section), multiplied by the Forecasted Sales for Future Cars. If the fuel-cell technology is selected in the Inputs section, the Total Direct Labor is the Direct Labor rate multiplied by the Direct Labor Hours for each fuel-cell car (both values are in the Constants section), multiplied by the Forecasted Sales for Future Cars.

- Total Manufacturing Overhead—This value is the total amount of money spent on manufacturing overhead to produce the forecasted number of cars. If the electric hybrid is selected in the Inputs section, the Total Manufacturing Overhead is the Direct Labor Hours required for each hybrid car multiplied by the Overhead Allocation Rate (both values are in the Constants section), multiplied by the Forecasted Sales for Future Cars Inc. If the fuel-cell technology is selected, the Total Manufacturing Overhead is the Direct Labor Hours required for each fuel-cell car multiplied by the Overhead Allocation Rate (both values are in the Constants section), multiplied by the Forecasted Sales for Future Cars Inc.

- Cost of Goods Sold—This value is the sum of the Total Direct Materials, Total Direct Labor, and Total Manufacturing Overhead required to produce the cars.

- Manufacturing Cost per Unit—This value is the Cost of Goods Sold divided by the Forecasted Sales for Future Cars Inc. This data is not required, but managers frequently want to see the actual cost of manufacturing for their products so they can try to reduce costs.
- Unit Sales Price—This value is the manufacturer's suggested retail price (MSRP). Marketing has developed a formula to set this price according to the following situations:

  - If the economic outlook is for a recession and the gasoline price is low, the sales price will be the same as the previous year's price (no increase).
  - If the economic outlook is for a recession and the gasoline price is high, the sales price will be two percent higher than the previous year's sales price (in other words, the previous year's sales price multiplied by 102 percent).
  - If the economic outlook is for a boom and the gasoline price is low, the sales price will be four percent higher than the previous year's sales price.
  - If the economic outlook is for a boom and the gasoline price is high, the sales price will be six percent higher than the previous year's sales price.

    (Hint: To set these prices, you will have to use a nested IF statement that contains AND functions.)

- Period Costs—These values are all the other nonmanufacturing costs incurred with running the new technology, including product improvement, marketing, distribution, and customer service costs. In this case, the Period Costs are roughly the same regardless of which technology is chosen, and are not dependent on the volume of cars produced. However, the costs do depend on the Economic Outlook (from the Inputs section). If the Economic Outlook is for a recession, the Period Costs are three percent higher than the previous year's Period Costs. If the Economic Outlook is for a boom, the Period Costs are six percent higher than the previous year's Period Costs.   C16 = R .03 × P pc   B = .06 × P pc

The Calculations section includes several complicated formulas. If you get lost while trying to write the nested IF and AND formulas, refer back to Tutorial C or ask your instructor for help.

Income and Cash Flow Statements Section

The statements for income and cash flow start with the cash on hand at the beginning of the year. Because Future Cars Inc. is funding the capital investment *internally*—that is, with its own cash on hand—you will have to deduct the invested funds from the cash on hand at the end of the year 2012. Figure 6-5 and the following list show how you should structure the Income and Cash Flow Statements section.

	A	B	C	D	E
37	**Income and Cash Flow Statements**	**2012**	**2013**	**2014**	**2015**
38	Beginning-of-year Cash on Hand (deduct Investment for 2013)	NA			
39	Total Sales Revenues	NA			
40	less: Cost of Goods Sold	NA			
41	Gross Profit	NA			
42	less: Period Costs	NA			
43	Net Profit before Income Tax	NA			
44	less: Income Tax Expense	NA			
45	Net Income after Taxes (Cash Inflow)	NA			
46	End-of-year Cash on Hand (Available to Invest)	$1,300,000,000			

FIGURE 6-5 Income and Cash Flow Statements section

- Beginning-of-year Cash on Hand—For 2013, this value is the End-of-year Cash on Hand from 2012 minus the capital investment, depending on the technology chosen. If you choose the hybrid electric technology, the capital investment will be $600 million (cell B10 in the Constants section). If you choose the fuel-cell technology, the capital investment will be $1.1 billion (cell B11 in the Constants section). For 2014 and 2015, the Beginning-of-year Cash on Hand will be the End-of-year Cash on Hand from the previous year.

- Total Sales Revenues—This value is the Forecasted Sales for Future Cars multiplied by the Unit Sales Price (both taken from the Calculations section).
- Less: Cost of Goods Sold—This value is the Cost of Goods Sold copied from the Calculations section.
- Gross Profit—This value is the Total Sales Revenues minus the Cost of Goods Sold.
- Less: Period Costs—These values are the Period Costs copied from the Calculations section.
- Net Profit before Income Tax—This value is the Gross Profit minus the Period Costs.
- Less: Income Tax Expense—If you make a profit (in other words, if the Net Profit before Income Tax is greater than zero), this value is the Net Profit before Income Tax multiplied by the Corporate Income Tax Rate from the Constants section. If you make nothing or have a net loss, the Income Tax Expense is zero.
- Net Income after Taxes (Cash Inflow)—This value is the Net Profit before Income Tax minus the Income Tax Expense. From a strict accounting standpoint, the Net Income after Taxes is not the cash inflow, because you would have to add back all noncash expenses such as depreciation and/or depletion to determine the true cash inflow. However, for the purposes of this case, assume that Net Income after Taxes is equal to cash inflow.
- End-of-year Cash on Hand—This value is the Beginning-of-year Cash on Hand plus the Net Income after Taxes.

Internal Rate of Return Calculation Section

This section, as shown in Figure 6-6, is set up to facilitate using Excel's built-in IRR (Internal Rate of Return) function.

	A	B
48	**Internal Rate of Return Calculation**	
49	Investment (Cash Outflow)	
50	Cash Inflow 2013	
51	Cash Inflow 2014	
52	Cash Inflow 2015	
53	Internal Rate of Return (IRR)	

FIGURE 6-6 Internal Rate of Return Calculation section

The IRR Calculation section includes the following values:

- Investment (Cash Outflow)—This value is the investment amount from cell B26 in the Calculations section multiplied by –1. This investment value must be a *negative* number to represent it as a Cash Outflow. (Think of it as money out of your pocket.)
- Cash Inflow 2013—This value is the Net Income After Taxes for 2013.
- Cash Inflow 2014—This value is the Net Income After Taxes for 2014.
- Cash Inflow 2015—This value is the Net Income After Taxes for 2015.
- Internal Rate of Return (IRR)—This value is the annual rate of return that your project is generating for the company. Many companies set a minimum IRR required for a project or investment to be selected for implementation. To calculate the IRR, click cell B53 (where you want the IRR result), then click the *fx* symbol next to the cell-editing window. The Insert Function dialog box appears (see Figure 6-7). Type IRR in the "Search for a function" text box and then click Go.

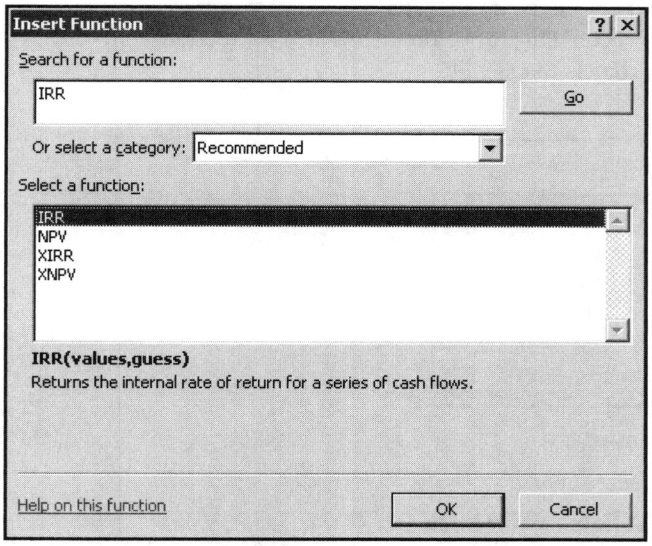

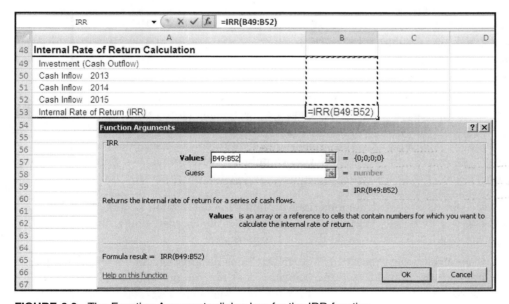

FIGURE 6-7 The Insert Function dialog box

The Function Arguments dialog box appears to help you build the formula (see Figure 6-8). In the Values text box, either enter the cells that contain all your cash outflows and inflows (B49:B52), or click and drag your mouse to select cells B49 through B52. Notice that Excel writes the formula for you in cell B53: =IRR(B49:B52). You do not have to enter a value in the Guess box. When you click OK, Excel calculates the IRR and places the result in cell B53.

FIGURE 6-8 The Function Arguments dialog box for the IRR function

After you complete all the formulas in the sections, try testing your spreadsheet with various combinations of the three inputs (there are eight possible combinations, as listed in the next section). If you receive any error messages or see strange values in the cells, go back and check your formulas.

The DSS spreadsheet model contains large numbers that represent millions or billions of dollars. Accountants often simplify their spreadsheets by listing the outputs in multiples of thousands or millions of dollars. It is not hard to do—you simply divide the cells by a thousand or a million depending on the scale—but for the purposes of this case, you should keep the large numbers in the spreadsheet. If you see cell results listed as a group of "#" signs when working with large numbers (see Figure 6-9), the cell is not wide enough to display the number—simply widen the column until the number is displayed.

	A	B	C
		2012	2013
37	Income and Cash Flow Statements		
38	Beginning-of-year Cash on Hand (deduct Investment for 2013)	NA	#########
39	Total Sales Revenues	NA	#########
40	less: Cost of Goods Sold	NA	#########
41	Gross Profit	NA	#########
42	less: Period Costs	NA	#########
43	Net Profit before Income Tax	NA	#########
44	less: Income Tax Expense	NA	$70,820,000
45	Net Income after Taxes (Cash Inflow)	NA	#########
46	End-of-year Cash on Hand (Available to Invest)	$1,300,000,000	#########

FIGURE 6-9   Column C is not wide enough to display the numbers

ASSIGNMENT 2: USING THE SPREADSHEET FOR DECISION SUPPORT

Next, you will use the spreadsheet to gather data needed to determine the best investment decision and to document your recommendations in a report to Future Cars Inc. management.

As stated before, this DSS model has eight possible financial outcomes:

1. Hybrid electric technology (E)
   a. Recession and low gasoline price (R/L/E)
   b. Recession and high gasoline price (R/H/E)
   c. Boom and low gasoline price (B/L/E)
   d. Boom and high gasoline price (B/H/E)
2. Fuel-cell technology (F)
   a. Recession and low gasoline price (R/L/F)
   b. Recession and high gasoline price (R/H/F)
   c. Boom and low gasoline price (B/L/F)
   d. Boom and high gasoline price (B/H/F)

You are primarily interested in the company's financial position based on each of the preceding possible outcomes. The Summary of Key Results section lets you see the Net Income after Taxes for each of the first three years of the new product sales, the End-of-year Cash on hand at the end of each of the three years, and the Internal Rate of Return. The management team wants to make sure that the selected technology does not exhaust the company's cash on hand at the end of any year examined (2013 through 2015). In addition, the team would prefer the technology that provides the best chance of making the targeted Internal Rate of Return of 20 percent or greater on the investment.

Because there are only eight (2^3) possible combinations of inputs for the economic outlook, gasoline prices, and technology selected, you might want to run the spreadsheet model eight times, changing the inputs according to the preceding list. You should do this for two reasons:

1. To ensure that no single year from 2013 through 2015 has negative income or end-of-year cash on hand (in other words, to make sure the company does not run out of money)
2. To print each spreadsheet to meet the requirements of Assignment 2A

You could then transcribe the results to a summary sheet. Next, you know that the management team is very interested in the financial data from the end of the third year of the model (2015). You can summarize that data easily using Scenario Manager.

Assignment 2A: Using Scenario Manager to Gather Data

For each of the eight situations listed earlier, you want to know the net income after taxes and the end-of-year cash on hand for the third year (2015) of the project, as well as the internal rate of return generated by the three years' cash inflows.

You will run "what-if" scenarios with the eight sets of input values using Excel Scenario Manager. If necessary, review Tutorial C for tips on using Scenario Manager. In this case, the input values are stored together in one vertical group of cells (C16 through C18) in the Inputs section, as are the three output cells (E21 through E23) in the Summary of Key Results section, so selecting them will be easy. Run Scenario Manager to gather your data in a report called the Scenario Summary. Format this summary to make it presentable, and then print it for your instructor.

If you have not done so already, you should run the spreadsheet model eight times, once for each input combination. Save each spreadsheet in your Excel workbook with a different sheet name. Keep these names as short but descriptive as possible; for example, you can use RLE, RHE, BLE, BHE, RLF, RHF, BLF, and BHF. Print the spreadsheets if your instructor requires it. Make sure to save your completed Excel workbook before closing it.

Assignment 2B: Documenting Your Recommendations in a Report

Use Microsoft Word to write a brief report to the CEO of Future Cars Inc., Ellen Coleman. State the results of your analysis and recommend which investment choice to make (electric hybrid or fuel cell). Your report must meet the following requirements:

- The first paragraph must summarize the technology choices facing Future Cars Inc. and state the purpose of the analysis.
- Next, summarize the results of your analysis and state your recommended action.
- Support your recommendation with a table outlining the Scenario Summary results. Figure 6-10 has a recommended Microsoft Word table format.
- If it is well formatted, you might choose to embed an Excel object of the Scenario Summary into the body of your report. Tutorial C includes a brief description of how to copy and paste Excel objects.
- Your instructor might also ask you to provide a graph of the internal rates of return for the eight possible combinations of inputs.

Technology	Economic Outlook	Gasoline Prices	2015 Net Income ($ millions)	2015 End-of-year Cash on Hand ($ millions)	Internal Rate of Return
Hybrid Electric	Recession	Low			
		High			
	Boom	Low			
		High			
Fuel Cell	Recession	Low			
		High			
	Boom	Low			
		High			

FIGURE 6-10   Recommended format of a table to insert in your report

In Figure 6-10, divide each of your Net Income and Cash on Hand results by 1 million to display them in multiples of millions.

ASSIGNMENT 3: GIVING A PRESENTATION WITH SLIDES

Your instructor may ask you to summarize your analysis in an oral presentation. If so, assume that the management team at Future Cars Inc. wants you to explain your analysis and recommendations in 10 minutes or less. A well-designed PowerPoint presentation, with or without handouts, is considered appropriate in a business setting. Tutorial F provides excellent tips for preparing and delivering a presentation.

DELIVERABLES

Your completed case should include the following deliverables for the instructor:

1. A printed copy of your report to management
2. Printouts of your spreadsheets
3. Electronic copies of all your work, including your report, PowerPoint presentation, and Excel DSS model. Ask your instructor for guidance on which items you should submit for grading.

THE HEALTH CARE COVERAGE DECISION AT BIG DOG COLLARS

Decision Support Using Excel

PREVIEW

Your dog collar business is doing well. You are considering adding health care coverage as an employee benefit. However, this coverage would be expensive, and other costs are expected to rise in the future. In this case, you will use Excel to see if the business will be profitable enough to allow adding employee health coverage.

PREPARATION

- Review spreadsheet concepts discussed in class and in your textbook.
- Complete any exercises that your instructor assigns.
- Complete any part of Tutorial C that your instructor assigns. You may need to review the use of IF statements and the section called "Cash Flow Calculations: Borrowing and Repayments."
- Review file-saving procedures for Windows programs.
- Refer to Tutorial E as necessary.

BACKGROUND

Your business is called Big Dog Collars. You make high-end dog collars and sell them directly to dog owners via the Internet.

You might think that dog owners would be happy to have a single collar for their pets, but you would be mistaken. Today's dog can have a collar for each season. The fall collar might feature a pattern of turning leaves. The winter collar might show a jolly Frosty the Snowman. Furthermore, owners can buy collars for special days, such as a July 4th collar that features the American flag. The marketing possibilities are almost endless.

Your dog collars have two parts. The strap that goes around the dog's neck is made from an attractive, durable fabric. The ends of the strap fit into a two-part plastic snap that secures the collar.

You buy the strap fabric and snaps from different suppliers. The strap fabric comes in a flat sheet that must be cut, folded, stitched, and ironed by one of your assembly workers. The snap has a male part and a female part. One end of the strap is fastened to the male part of the snap, and the other end is fastened to the female part.

The assembly process sounds simple, but it is exacting work. The cutting, folding, and stitching must be precise. The strap ends must be inserted into the snaps without crinkling the fabric. You have four good manufacturing workers whom you pay well so that they stay with your company. Each worker can make 20 collars per day, and they work 250 days per year.

It is amazing how many products contain petroleum as a raw material, and the collar parts are no exception. Oil is refined into compounds that become feedstocks in the manufacture of many products, including plastics and synthetic fibers.

The companies that make your snaps and fabric pass along changes in their costs, meaning that a change in the price of petroleum will affect the cost of the snaps and fabric. In recent years, world oil prices have fluctuated greatly. Many economists and petroleum experts think that the price of oil will increase significantly in this decade, which makes you nervous about cost control in the next few years. On the other hand, some experts think that oil prices will not change much, and might even decline.

The price of oil is commonly quoted by the barrel. In 2011, the average price of a barrel of oil on the open market was $80. Big Dog's petroleum-based raw material costs are about 50% of the percentage change in the price of a barrel of oil. For example, if the price of oil rose from $80 in one year to $90 in the next year, the increase would be 12.5% ((90 − 80) / 80 = .125). Therefore, you would expect Big Dog's raw material costs to increase by 6.25% (.5 × .125) in the second year.

Even in poor economic times, people are willing to pay for a good specialty product like your dog collar. Your advertising on the Internet, other advertising, and word of mouth seem to ensure that your sales will continue to increase. You think that unit sales will increase 3% each year for the next three years. Likewise, you think that you can raise your selling price 3% each year for the next three years. Increasing production may require adding one or more manufacturing workers.

You also employ an office worker and a janitor/repairman. If you adopt health care as a benefit, you would cover all your employees. (You would not cover yourself because your wife is a public school teacher, and her health care plan covers you.) You have identified a basic three-year health care package that would cost you $3000 per covered employee in 2012, $3200 in 2013, and $3400 in 2014.

Each year, your financial goal is to achieve a 5% "return on sales"; in other words, you want net income after taxes divided by total revenue to be at least 5%. If the return is greater than 5%, you conclude that sales and costs were in control for the year. You can conclude the opposite if the return is less than 5%.

You have achieved more than a 5% return on sales in recent years. You have paid off most of the bank debt that you used to purchase manufacturing equipment and other fixtures. You have cash in the bank. Your banker says she would be willing to finance an expansion of the business if necessary, and that the bank could cover you in lean years.

Looking ahead, you certainly do not want good employees to quit because they are seeking health care coverage elsewhere. If you pay for health care and oil prices increase, can you still make your profit goal in the foreseeable future? If oil prices increase, can you make your profit goal *only if* you do not adopt health care? Covering your employees' health care would be the right thing to do, but making your profit goal is also important.

Your DSS model needs to account for changes in oil-based costs, with and without the adoption of employee health care. Your model will let you develop "what-if" scenarios with the inputs, see the financial results, and help you decide whether you should adopt health care as a benefit.

ASSIGNMENT 1: CREATING A SPREADSHEET FOR DECISION SUPPORT

In this assignment, you will produce a spreadsheet that models the business decision. Then, in Assignment 2, you will write a memorandum that documents your analysis and conclusions. In Assignment 3, you will prepare and give a presentation of your analysis and conclusions to your banker.

First, you will create the spreadsheet model of the financial situation. The model covers the three years from 2012 to 2014. This section helps you set up each of the following spreadsheet components before entering cell formulas:

- Constants
- Inputs
- Summary of Key Results
- Calculations
- Income and Cash Flow Statements
- Debt Owed
- Return on Sales

A discussion of each section follows. The spreadsheet skeleton is available for you to use. To access the spreadsheet skeleton, go to your data files, select Case 7, and then select **DogCollar.xlsx**.

Constants Section

Your spreadsheet should include the constants shown in Figure 7-1. An explanation of the line items follows the figure.

	A	B	C	D	E
1	**Health Care Coverage at Big Dog Collars**				
2					
3	**Constants**	**2011**	**2012**	**2013**	**2014**
4	Tax Rate	NA	30%	32%	34%
5	Minimum cash needed to start year	NA	$25,000	$25,000	$25,000
6	Fixed Costs	NA	$100,000	$110,000	$120,000
7	Health care cost per employee	NA	$3,000	$3,200	$3,400
8	Interest rate	NA	4%	5%	6%

FIGURE 7-1   Constants section

- Tax Rate—The tax rate is applied to income before taxes. The rate is expected to increase each year.
- Minimum cash needed to start year—You want to have at least $25,000 in cash at the beginning of each year. Your banker will lend you the amount you need at the end of a year in order to begin the new year with $25,000.
- Fixed Costs—These costs include rent, maintenance, insurance, electricity, and salaries for the office worker and handyman. These costs do not fluctuate with changes in production. They are expected to increase each year.
- Health care cost per employee—The insurance company will charge $3,000 per covered employee in 2012, and more in succeeding years.
- Interest rate—Your bank charges interest on any borrowing. Your banker says that interest rates are expected to rise as the economy continues to recover.

Inputs Section

Your spreadsheet should include the following inputs for the years 2012, 2013, and 2014, as shown in Figure 7-2.

	A	B	C	D	E
10	**Inputs**	**2011**	**2012**	**2013**	**2014**
11	Expected cost of a barrel of oil	NA			
12	Cover employee health care? (YES/NO)		NA	NA	NA

FIGURE 7-2   Inputs section

- Expected cost of a barrel of oil—Enter the expected average cost of a barrel of oil in each year. For example, you could enter 90, 100, and 110 to indicate a gradually increasing cost of oil. The pattern 80, 70, 60 would indicate a gradually decreasing cost.
- Cover employee health care? (YES/NO)—Enter YES if employee health care will be paid for in 2012, 2013, and 2014. Otherwise, enter NO. The single entry applies to all three years.

Summary of Key Results Section

Your spreadsheet should include the results shown in Figure 7-3.

	A	B	C	D	E
14	**Summary of Key Results**	**2011**	**2012**	**2013**	**2014**
15	Net income after taxes	NA			
16	End-of-year cash on hand	NA			
17	End-of-year debt owed	NA			
18	Return on sales	NA			

FIGURE 7-3   Summary of Key Results section

For each year, your spreadsheet should show net income after taxes, cash on hand at the end of the year, bank debt owed at the end of the year, and return on sales for the year. Format the net income, cash, and debt cells as currency with zero decimals. Format the return on sales cells as a percentage with one decimal place. These values are computed elsewhere in the spreadsheet and should be echoed here.

Calculations Section

You should calculate intermediate results to use in the income and cash flow statements that follow. Calculations, as shown in Figure 7-4, may be based on expected year-end 2011 values. When called for, use absolute referencing properly. You must compute values by cell formula; use hard-coded numbers in formulas only when instructed. Cell formulas should not reference a cell that contains a value of "NA."

Big Dog makes collars for dogs of all sizes. The product mix in 2012–2014 is expected to stay the same; in other words, the proportion of small collars to all collars will stay the same, the proportion of large collars will stay the same, and so on. Thus, prices and costs calculated here are averages for the company's product mix of sizes sold.

An explanation of each item in this section follows the figure.

	A	B	C	D	E
20	**Calculations**	**2011**	**2012**	**2013**	**2014**
21	Selling price	$ 20.00			
22	Units sold	20,000			
23	Oil cost change factor	NA			
24	Raw material cost change factor	NA			
25	Snap cost	$ 0.50			
26	Strap cost	$ 1.00			
27	Number of manufacturing employees	4			
28	Manufacturing employee salary	$ 50,000			

FIGURE 7-4   Calculations section

- Selling price—The average selling price for collars of all sizes was $20 in 2011. The mix of sizes sold is expected to stay the same in the future. The average selling price is expected to increase 3% each year. In other words, the 2012 value will be 3% greater than the 2011 value, the 2013 value will be 3% greater than the 2012 value, and so on. Format cells as currency with two decimal places.
- Units sold—Big Dog sold a total of 20,000 collars in 2011. The mix of sizes sold is expected to stay the same in the future. The number of collars sold in a year is expected to increase 3% each year. Format these cells as a number with no decimal places.
- Oil cost change factor—This value is the percentage change in the average price of a barrel of oil from one year to the next. Note that the price was $80 in 2011; this number can be hard-coded into the 2012 formula. Format this value as a percentage with one decimal place.
- Raw material cost change factor—This factor is half of the year's oil cost change factor. Format this factor as a percentage with one decimal place.
- Snap cost—This value is the expected average cost of collar snaps per year. The cost is based on the prior year's cost, and is increased or decreased by the raw material cost change factor. For example, if the snap cost is $0.50 in one year and the raw material cost change factor is 10% the next year, the expected snap cost in the next year would be $.50 × (1 + 0.10) = $0.55. Format this value as currency with two decimal places.
- Strap cost—This value is the expected average cost of collar straps per year. The cost computation is the same as the snap cost calculation. Format the value as currency with two decimal places.
- Number of manufacturing employees—A manufacturing employee assembles 5,000 collars per year (20 per day × 250 work days). To determine how many manufacturing employees you need, divide the expected production by 5,000 using the Roundup function. For example, if cell M20 holds the number 102, the formula =Roundup(M20/5,0) would return the value 21 with no decimal places.

- Manufacturing employee salary—The average manufacturing employee salary is expected to increase 2% per year in each of the next three years. Format the value as currency with no decimal places.

Income and Cash Flow Statements

The forecast for net income and cash flow starts with the cash on hand at the beginning of the year, continues with the income statement, and concludes with the calculation of cash on hand at year's end. For readability, format cells in this section as currency with zero decimal places. You must compute values by cell formula; use hard-coded numbers in formulas only when instructed. Cell formulas should not reference a cell that contains a value of "NA." Your spreadsheets should look like those shown in Figures 7-5 and 7-6. A discussion of each item in the section follows each figure.

	A	B	C	D	E
30	**Income and Cash Flow Statements**	**2011**	**2012**	**2013**	**2014**
31	Beginning-of-year cash on hand	NA			
32					
33	Revenue	NA			
34	Costs:	NA	-	-	-
35	Manufacturing employee salary	NA			
36	Raw material costs	NA			
37	Fixed costs	NA			
38	Health care cost	NA			
39	Total Costs	NA			
40	Income before interest and taxes	NA			
41	Interest expense	NA			
42	Income before taxes	NA			
43	Income tax expense	NA			
44	Net Income after taxes	NA			

FIGURE 7-5 Income and Cash Flow Statements section

- Beginning-of-year cash on hand—This amount is the cash on hand at the end of the prior year.
- Revenue—This amount is a function of the expected units sold in the year and the average selling price for the year. Both of these values are taken from the Calculations section.
- Manufacturing employee salary—This amount is a function of the number of manufacturing employees in the year and the average manufacturing employee salary expected in the year. Both of these values are taken from the Calculations section.
- Raw material costs—This amount is a function of the expected units sold in the year and the expected snap costs and strap costs in the year. All of these values are taken from the Calculations section.
- Fixed costs—This amount is a constant that can be echoed here.
- Health care cost—This amount is a function of the year's health care cost per employee (a constant) and the total number of employees to be covered.
- Total Costs—This amount is the total of manufacturing salaries, raw material costs, fixed costs, and health care costs.
- Income before interest and taxes—This amount is the difference between revenue and total costs.
- Interest expense—This amount is the product of the debt owed to the bank at the beginning of the year and the interest rate for the year (a constant).
- Income before taxes—This amount is income before interest and taxes minus interest expense.
- Income tax expense—This amount is zero if income before taxes is zero or less. Otherwise, income tax expense is the product of the year's tax rate (a constant) and income before taxes.
- Net income after taxes—This amount is the difference between income before taxes and income tax expense.

Line items for the year-end cash calculation are discussed next. In Figure 7-6, column B represents 2011, column C is for 2012, and so on. Year 2011 values are NA, except for End-of-year cash on hand, which is $25,000.

	A	B	C	D	E
46	Net Cash Position (NCP)	NA			
47	Borrowing from bank	NA			
48	Repayment to bank	NA			
49	End-of-year cash on hand	$ 25,000			

FIGURE 7-6   End-of-year cash on hand section

- Net Cash Position (NCP)—The NCP at the end of a year equals the cash at the beginning of the year plus the year's net income after taxes.
- Borrowing from bank—Assume that the company's bank will lend enough money at the end of the year to reach the minimum cash needed to start the next year. If the NCP is less than this minimum, the company must borrow enough to start the next year with the minimum. Borrowing increases the cash on hand, of course.
- Repayment to bank—If the NCP is more than the minimum cash needed and some debt is owed at the beginning of the year, you must pay off as much debt as possible without going below the minimum cash required to start the next year. Repayments reduce cash on hand.
- End-of-year cash on hand—This amount is the NCP plus any borrowing and minus any repayments.

Debt Owed Section

This section shows a calculation of debt owed to the bank at year's end (see Figure 7-7). Year 2011 values are NA, except for End-of-year debt owed, which is $50,000. You must compute values by cell formula; use hard-coded numbers in formulas only when instructed. Cell formulas should not reference a cell that contains a value of "NA." An explanation of each item follows the figure.

	A	B	C	D	E
51	**Debt Owed**	**2011**	**2012**	**2013**	**2014**
52	Beginning-of-year debt owed	NA			
53	Borrowing from bank	NA			
54	Repayment to bank	NA			
55	End-of-year debt owed	$ 50,000			

FIGURE 7-7   Debt Owed section

- Beginning-of-year debt owed—Debt owed at the beginning of a year equals the debt owed at the end of the prior year.
- Borrowing from bank—This amount has been calculated elsewhere and can be echoed to this section. Borrowing increases the amount of debt owed.
- Repayment to bank—This amount has been calculated elsewhere and can be echoed to this section. Repayments reduce the amount of debt owed.
- End-of-year debt owed—In 2012 through 2014, this value is the amount owed at the beginning of a year, plus borrowing during the year, minus repayments during the year.

Return on Sales Section

This section shows a calculation of return on sales in each year. Return on sales is net income after taxes for the year divided by revenue for the year, as shown in Figure 7-8. Format values as percentages with one decimal place.

	A	B	C	D	E
57	**Return on Sales**	**2011**	**2012**	**2013**	**2014**
58	Net Income after taxes / Revenue	NA			

FIGURE 7-8   Return on Sales section

Recall that the company's goal is to earn at least 5% return on sales each year.

ASSIGNMENT 2: USING THE SPREADSHEET FOR DECISION SUPPORT

Complete the case by using the spreadsheet to determine your health care coverage strategy and by documenting recommendations in a memorandum.

You want to model four oil price situations, which you call Probable, Possible, Worst Case, and Optimistic:

- In the Probable situation, per-barrel oil prices escalate slightly in 2012–2014 to $100, $110, and $120, respectively.
- In the Possible situation, per-barrel oil prices escalate significantly in 2012–2014 to $100, $125, and $150, respectively.
- In the Worst Case situation, per-barrel oil prices escalate greatly in 2012–2014 to $100, $150, and $200, respectively.
- In the Optimistic situation, per-barrel oil prices actually decrease in 2012–2014 to $100, $90, and $80, respectively.

Within each of these four situations, you can either choose to pay or not pay for employee health care coverage. The added choice results in the following eight scenarios:

1. Prob-No—This scenario uses the Probable oil price situation and assumes that you are not paying for health care.
2. Prob-Yes—This scenario uses the Probable oil price situation and assumes that you are paying for health care.
3. Poss-No—This scenario uses the Possible oil price situation and assumes that you are not paying for health care.
4. Poss-Yes—This scenario uses the Possible oil price situation and assumes that you are paying for health care.
5. Worst-No—This scenario uses the Worst Case oil price situation and assumes that you are not paying for health care.
6. Worst-Yes—This scenario uses the Worst Case oil price situation and assumes that you are paying for health care.
7. Opt-No—This scenario uses the Optimistic oil price situation and assumes that you are not paying for health care.
8. Opt-Yes—This scenario uses the Optimistic oil price situation and assumes that you are paying for health care.

You want answers to the following questions:

- If you pay for health care and oil prices increase, can you still make your profit goal in the foreseeable future?
- If oil prices increase, can you make your profit goal *only if* you do not adopt health care?
- How much impact do oil prices actually have for you? Looking at 2014 net income, cash, debt, and return on sales, how different are the Optimistic-Yes and Probable-Yes results? For that matter, how much different are the Optimistic-Yes and Worst-Yes results?
- Is there a case to be made for paying for health care in 2012–2014, no matter what? Can the case be made in terms of 2014 net income, cash, debt, and return on sales?

You will use your spreadsheet to gather data about these questions.

Assignment 2A: Using the Spreadsheet to Gather Data

You have built the spreadsheet to model the business situation. For each of the eight scenarios, you want to know the 2014 net income after taxes, the 2014 end-of-year cash on hand, the 2014 end-of-year debt owed, and the 2014 return on sales.

You will run "what-if" scenarios with the eight sets of input values using Scenario Manager. (See Tutorial C for details on using Scenario Manager.) Set up the eight scenarios. Your instructor may ask you to use conditional formatting to make sure that your input values are proper. (Note that you can enter noncontiguous cell ranges in Scenario Manager, such as C19, D19, C20:F20.)

The relevant output cells are the four 2014 cells in the Summary of Key Results section. Run Scenario Manager to gather the data in a report. When you finish, print the spreadsheet with the input for any one of the scenarios, print the Scenario Manager summary sheet, and then save the spreadsheet file for the last time.

Assignment 2B: Documenting Your Recommendations in a Memorandum

Use Microsoft Word to write a brief memorandum documenting your analysis and conclusions. You can address the memo to your banker. Observe the following requirements:

- Set up your memo as described in Tutorial E.
- In the first paragraph, briefly state the business situation and the purpose of your analysis.
- Next, provide the answers to your four questions.
- State your recommendation: should the company provide health insurance?
- Support your statements graphically, as your instructor requires. Your instructor may ask you to return to Excel and copy the Scenario Manager summary sheet results into the memo. (See Tutorial C for details on this procedure.) Your instructor might also ask you to create a summary table in Word based on the Scenario Manager summary sheet results. (This procedure is described in Tutorial E.)

Your table should resemble the format shown in Figure 7-9.

Scenario	2014 Net Income	2014 Cash on Hand	2014 Debt Owed	2014 Return on Sales
Prob-No				
Prob-Yes				
Poss-No				
Poss-Yes				
Worst-No				
Worst-Yes				
Opt-No				
Opt-Yes				

FIGURE 7-9   Format of table to insert in memo

ASSIGNMENT 3: GIVING AN ORAL PRESENTATION

Your instructor may ask you to summarize your analysis and recommendations in an oral presentation. If so, assume that your banker wants the presentation to last 10 minutes or less. Use visual aids or handouts that you think are appropriate. See Tutorial F for tips on preparing and giving a presentation.

DELIVERABLES

Assemble the following deliverables for your instructor:

1. Printout of your memo
2. Spreadsheet printouts
3. Electronic media such as a USB key or CD that contains your Word memo and Excel spreadsheet file

Staple the printouts together with the memo on top. If you have more than one .xlsx file on your electronic media, write your instructor a note that identifies your model's .xlsx file.

PART 3

DECISION SUPPORT CASES USING THE EXCEL SOLVER

BUILDING A DECISION SUPPORT SYSTEM USING EXCEL SOLVER

In Tutorial C, you learned that Decision Support Systems (DSS) are programs used to help managers solve complex business problems. Cases 6 and 7 were DSS models that used Excel Scenario Manager to calculate and display financial outcomes given certain inputs, such as economic outlooks and mortgage interest rates. The outputs from Scenario Manager allowed you to see how different combinations of inputs affected cash flows and income so that you could make the best decision for expanding your business or selecting a technology to develop and market.

Many business situations require models in which the inputs are not limited to two or three choices, but include large ranges of numbers in more than three variables. For such business problems, managers want to know the best or optimal solution to the model. An optimal solution can either maximize an objective variable, such as income or revenues, or minimize the objective variable, such as operating costs. The formula or equation that represents target income or operating cost is called an objective function. Optimizing the objective function requires the use of constraints (also called constraint equations), which are rules or conditions you must observe when solving the problem. The field of applied mathematics that addresses problem solving with objective functions and constraint equations is called linear programming. Before the advent of digital computers, linear programming required the knowledge of complex mathematical techniques. Fortunately, Excel has a tool called Solver that can compute the answers to optimization problems.

This tutorial has five sections:

1. **Adding Excel Solver to the Excel Ribbon**—Solver is not installed with Excel 2010; you must add it to the application. You may need to use Excel Options to add Solver to the Ribbon.
2. **Using Excel Solver**—This section explains how to use Solver. You will start by using Solver to determine the best mix of vehicles for shipping exercise equipment to stores throughout the country.
3. **Extending the example**—This section tests your knowledge of Solver as you modify the transportation mix to accommodate changes: additional new stores to supply and redesign of the product to reduce shipping volume.
4. **Using Solver on a new problem**—In this section, you will use Solver on a new problem: maximizing the profits for a mix of products.
5. **Troubleshooting Solver**—Because Solver is a complex tool, you will sometimes have problems using it. This section explains how to recognize and overcome such problems.

NOTE

If you need a refresher, Tutorial C offers guidance on basic Excel concepts such as formatting cells and using the =IF( ) and AND( ) functions.

ADDING EXCEL SOLVER TO THE EXCEL RIBBON

Before you can use Excel Solver, you must determine whether it is installed in Excel. Start Excel and then click the Data tab in the Ribbon. If you see a group on the right side named Analysis that contains Solver, you do not need to install Solver (see Figure D-1).

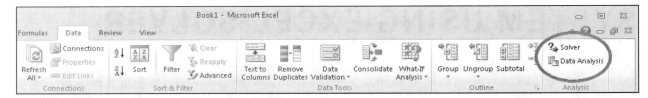

FIGURE D-1   Analysis group with Solver installed

If the Analysis group or Solver is not shown on the Data tab of the Ribbon, do the following:

1. Click the File tab.
2. Click Options (see Figure D-2).
3. Click Add-Ins (see Figure D-3) to display the available add-ins in the right pane.
4. Click Go at the bottom of the right pane. The dialog box shown in Figure D-4 appears.
5. Click the Solver Add-in box as well as the Analysis ToolPak and Analysis ToolPak-VBA boxes. (You will need the latter options in a subsequent case, so install them now with Solver.)
6. Click OK to close the window and return to the Ribbon. If you click the Data tab again, you should see the Analysis group with Data Analysis and Solver on the right.

FIGURE D-2   Excel Options selection

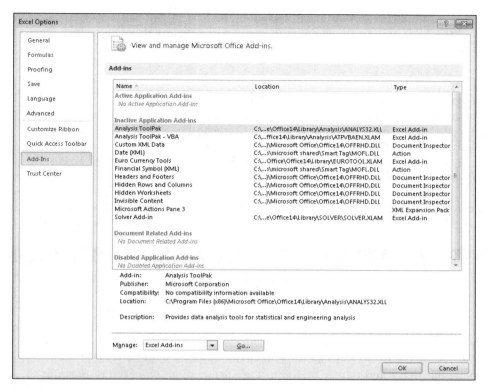

FIGURE D-3 Add-Ins pane

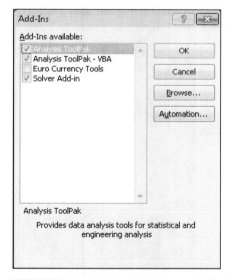

FIGURE D-4 Add-Ins dialog box with Solver, Analysis ToolPak, and Analysis ToolPak VBA selected

USING EXCEL SOLVER

A fictional company called CV Fitness builds exercise machines in its plant in Memphis, Tennessee and ships them to its stores across the country. The company has a small fleet of trucks and tractor-trailers to ship its products from the factory to its stores. It costs less money per cubic foot of capacity to ship products with tractor-trailers than with trucks, but the company has a limited number of both types of vehicles, and must ship a specified amount of each type of product to each destination. You have been asked to determine the optimal mix of trucks and tractor-trailers to send merchandise to each store. The optimal mix will have the lowest total shipping cost while ensuring that the required quantity of products is shipped to each store.

To use Solver, you must set up a model of the problem, including the factors that can vary (the mix of trucks and tractor-trailers) and the constraints on how much they can vary (the number of each vehicle available). Your goal is to minimize the shipping cost; you will execute Solver to compute the best solution.

Setting Up a Spreadsheet Skeleton

CV Fitness makes three fitness machines: exercise bikes (EB), elliptical cross-trainers (CT), and treadmills (TM). When packaged for shipment, their shipping volumes are 12, 15, and 22 cubic feet, respectively. The finished machines are shipped via ground transportation to five stores in Philadelphia, Atlanta, Miami, Chicago, and Los Angeles. Your vehicle fleet consists of 12 trucks and six tractor-trailers. Each truck has a capacity of 1500 cubic feet, and each tractor-trailer has a capacity of 2350 cubic feet. The spreadsheet includes the road distances from your plant in Memphis to each store, and each store's demand for the three fitness machines.

What is the best mix of trucks and tractor-trailers to send to each destination? You will learn how to use Solver to determine the answer. The spreadsheet components are discussed in the following sections.

AT THE KEYBOARD

Start by saving your blank spreadsheet as **CV Fitness.xlsx**. Then enter the skeleton and formulas as directed in the following sections.

Spreadsheet Title

Resize Column A, as illustrated in Figure D-5, to give your spreadsheet a small border on the left side. Enter the spreadsheet title in cell B1. Assign the title a font of Calibri Bold and a font size of 14 points. Merge and center cells B1 through F1 using the Merge and Center button in the Alignment group of the Home tab.

Constants Section

Your spreadsheet should have a section for values that will not change. Figure D-5 shows a skeleton of the Constants section and the values you should enter. A discussion of the line items follows the figure.

	A	B	C	D	E	F
1		CV Fitness, Inc. Truck Load Management Problem				
2						
3		Constants Section:				
4			Volume Cu. Ft.	Operating Cost per mi.	Operating Cost per mi-cu. Ft.	Available Fleet
5		Truck	1500	$1.00	$0.000667	12
6		Tractor Trailer	2350	$1.30	$0.000553	6
7						
8		Exercise Bike (EB)	12			
9		Elliptical Crosstrainer (CT)	15			
10		Treadmill (TM)	22			

FIGURE D-5 Spreadsheet title and Constants section

- In column C, enter the Volume Cu. Ft., which is the cubic-foot capacity of the vehicles as well as the shipping volume for each item of exercise equipment.
- In column D, enter the Operating Cost per mi., which is the cost per mile driven for each type of vehicle.
- In column E, enter the Operating Cost per mi.-cu.ft. This value is actually a formula: the operating cost per mile divided by the vehicle volume in cubic feet. Normally you do not put formulas in the Constants section, but in this case it lets you see the relative cost efficiencies of each vehicle. Assuming that both types of vehicles can be filled to capacity, the tractor-trailer is the preferred vehicle for shipping cost efficiency.

- • In column F, enter the values for the Available Fleet, which is the number of each type of vehicle your company owns or leases.

You can update the Constants section as the company adds more products to its offerings or adds vehicles to its fleet.

> **NOTE**
>
> The column headings in the Constants section contain two or three lines to keep the columns from becoming too wide. To create a multiple-line column heading, hold down the Alt key and press Enter when you need to create a new line.

Now is a good time to save your workbook. Use a descriptive filename so you can find it easily later—**CV Fitness Trucking Problem.xlsx** should work well.

Calculations and Results Section

The structure and format of your Calculations and Results section will vary greatly depending on the nature of the problem you need to solve. In some Solver models, you might need to maximize income, which means you might also have an Income Statement section. In other Solver models, you may want to have a separate Changing Cells section that contains cells Solver will manipulate to obtain a solution. However, in this tutorial you want to minimize shipping costs while meeting the product demand of your stores. You can accomplish this task by building a single unified table that includes the distances to the stores, the product demand for each store, and the shipping alternatives and costs.

A unified Calculations and Results section makes sense in this model for several reasons. First, it simplifies writing and copying the formulas for the needed shipping volumes, the vehicle capacity totals, and the shipping costs to each destination. Second, a well-organized table allows you to easily identify the Changing Cells, which Solver will manipulate to optimize the solution, as well as the Total Cost (or Optimization Cell). Finally, a unified table allows your management team to visualize the problem and its solution.

When creating a complex table, it is often a good idea to sketch the table's structure first to see how you want to organize the data. Format the table structure, then enter the data you are given for the problem. Write the cells that contain the formulas last, starting with all the formulas in the first row. If you do a good job structuring your table, you will be able to copy the first-row formulas to the other rows.

Build the blank table shown in Figure D-6. A discussion of the rows and columns follows the figure.

	Distance/Demand Table		Store Demand			Vehicle Loading							Cost
	Distance Table (from Memphis Plant)	Miles	EB	CT	TM	Volume Required	Trucks	Volume for Trucks	Tractor-Trailers	Volume for Tractor-Trailers	Total Vehicle Capacity	% of Vehicle Capacity Utilized	Shipping Cost
Philadelphia Store		1010	140	96	86								
Atlanta Store		380	76	81	63								
Miami Store		1000	56	64	52								
Chicago Store		540	115	130	150								
Los Angeles Store		1810	150	135	180								
	Totals:												
Fill Legend:			Changing Cells										Total Cost
			Optimization Cell										

FIGURE D-6 Blank table for Calculations and Results section

- • In row 13, enter "Calculations and Results Section:" as the title of the table.
- • In row 14, columns B and C, enter "Distance/Demand Table" as a column heading. Merge and center the heading in the two columns.
- • In row 14, columns D, E, and F, enter "Store Demand" as a column heading. Merge and center the heading in the three columns.
- • In row 14, columns G through M, enter "Vehicle Loading" as a column heading. Merge and center the heading across the columns.
- • In row 14, column N, enter "Cost" as a centered column heading.
- • In row 15, column B, enter "Distance Table (from Memphis Plant)" as a column heading.
- • In row 15, column C, enter "Miles" as a column heading.

- In row 15, columns D, E, and F, enter "EB," "CT," and "TM," respectively, as equipment headings.
- In row 15, columns G through N, enter "Volume Required," "Trucks," "Volume for Trucks," "Tractor-Trailers," "Volume for Tractor-Trailers," "Total Vehicle Capacity," "% of Vehicle Capacity Utilized," and "Shipping Cost," respectively, as column headings.
- In rows 16 through 20, column B, enter the destination store locations.
- In rows 16 through 20, column C, enter the number of miles to the destination store locations.
- In rows 16 through 20, columns D through F, enter the number of exercise bikes (EB), cross-trainers (CT), and treadmills (TM) to be shipped to each store location.
- Rows 16 through 20, columns G through N, will contain formulas or "seed values" later. Leave them blank for now, but fill cells H16 through H20 and cells J16 through J20 with a light color to indicate that they are the Changing Cells for Solver.
- In cell F21, enter "Totals:" to label the following cells in the row.
- Cells G21 through N21 will be used for column totals. Highlight cell N21 with a different light color fill from the Changing Cells. Cell N21 is your Optimization Cell.
- In cell B22, enter "Fill Legend:" as a label.
- Fill cell C22 with the fill color you selected for the Changing Cells.
- In cells D22 and E22, enter "Changing Cells" as the label for the fill color. Merge and center the label in the cells.
- In cell N22, enter "Total Cost" as the label for the value in cell N21.
- Fill cell C23 with the fill color you selected for the Optimization Cell.
- In cells D23 and E23, enter "Optimization Cell" as the label for the fill color. Merge and center the label in the cells.

Figure D-7 illustrates a magnified section of the Distance/Demand table in case the numbers in Figure D-6 are difficult to read.

	A	B	C	D	E	F
12						
13		**Calculations and Results Section:**				
14		**Distance/Demand Table**		**Store Demand**		
15		**Distance Table (from Memphis Plant)**	**Miles**	**EB**	**CT**	**TM**
16		Philadelphia Store	1010	140	96	86
17		Atlanta Store	380	76	81	63
18		Miami Store	1000	56	64	52
19		Chicago Store	540	115	130	150
20		Los Angeles Store	1810	150	135	180
21						Totals:
22		Fill Legend:		Changing Cells		
23				Optimization Cell		

FIGURE D-7   Magnified view of the Distance/Demand table

Use the Borders menu in the Font group to select and place appropriate borders around parts of the Calculations and Results section (see Figure D-8). The All Borders and Outside Borders selections are the most useful borders for your table.

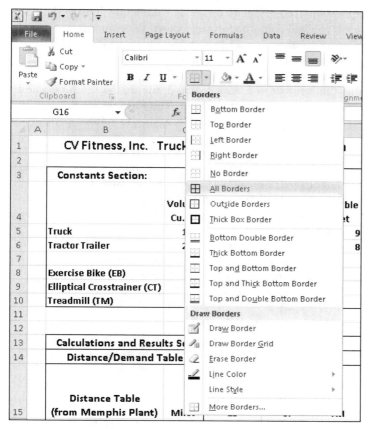

FIGURE D-8   Borders menu

Next, you write the formulas for the volume and cost calculations. Figure D-9 shows a magnified view of the Vehicle Loading and Cost sections. A discussion of the formulas required for the cells follows the figure.

	Vehicle Loading						Cost
Volume Required	Trucks	Volume for Trucks	Tractor-Trailers	Volume for Tractor-Trailers	Total Vehicle Capacity	% of Vehicle Capacity Utilized	Shipping Cost
							Total Cost

FIGURE D-9   Vehicle Loading and Cost sections

For illustration purposes, the cell numbers in the following list refer to values for the Philadelphia store.

- Volume Required—Cell G16 contains the total shipping volume of the three types of equipment shipped to the Philadelphia store. The formula for this cell is =D16*$C$8 + E16*$C$9 + F16*$C$10. Cells D16, E16, and F16 are the quantities of each item to be shipped, and cells $C$8, $C$9, and $C$10 are the shipping volumes for the exercise bike, cross-trainer, and treadmill, respectively.

When taking values from the Constants section to calculate formulas, you almost always should use absolute cell references ($) because you will copy the formulas down the columns.

- Trucks—Cell H16 contains the number of trucks selected to ship the merchandise. Cell H16 is a Changing Cell, which means Solver will determine the best number of trucks to use and place the number in this cell. For now, you should "seed" the cell with a value of 1.
- Volume for Trucks—Cell I16 contains the number of trucks selected, multiplied by the capacity of a truck. The capacity value is taken from the Constants section. The formula for this cell is =H16*$C$5. Cell H16 is the number of trucks selected, and cell $C$5 is the volume capacity of the truck in cubic feet.
- Tractor-Trailers—Cell J16 contains the number of tractor-trailers selected to ship the merchandise. Cell J16 is a Changing Cell, which means Solver will determine the best number of tractor-trailers to use and place the number in this cell. For now, you should "seed" the cell with a value of 1.
- Volume for Tractor-Trailers—Cell K16 contains the number of tractor-trailers selected, multiplied by the capacity of a tractor-trailer. The capacity value is taken from the Constants section. The formula for this cell is =J16*$C$6. Cell J16 is the number of tractor-trailers selected, and cell $C$6 is the cubic feet capacity of the tractor-trailer.
- Total Vehicle Capacity—Cell L16 contains the sum of the Volume for Trucks and the Volume for Tractor-Trailers. The formula for this cell is =I16+K16. You need to know the Total Vehicle Capacity to make sure that you have enough capacity to ship the Volume Required. This value will be one of your constraints in Solver.
- % of Vehicle Capacity Utilized—Cell M16 contains the Volume Required divided by the Total Vehicle Capacity. The formula for this cell is =G16/L16; after entering the formula, click the % button in the Number group. Although this information is not required to minimize shipping costs, it is useful for managers to know how much space was filled in the selected vehicles. Alternatively, you could run Solver to determine the highest space utilization on the vehicles rather than the lowest cost. Note that you cannot use more than 100% of the available space on the trucks.
- Shipping Cost—Cell N16 contains the following calculation:

  Mileage to destination store × Number of trucks selected × Cost per mile for trucks + Mileage to destination store × Number of tractor-trailers selected × Cost per mile for tractor-trailers

  The formula for this cell is =H16*C16*$D$5+J16*C16*$D$6. Note that absolute cell references for the cost-per-mile values are taken from the Constants section.

If you entered the formulas correctly in row 16, your table should look like Figure D-10.

	G	H	I	J	K	L	M	N
14				Vehicle Loading				Cost
15	Volume Required	Trucks	Volume for Trucks	Tractor-Trailers	Volume for Tractor-Trailers	Total Vehicle Capacity	% of Vehicle Capacity Utilized	Shipping Cost
16	5012	1	1500	1	2350	3850	130%	$2,323.00
17								
18								
19								
20								
21								
22								Total Cost

FIGURE D-10 Vehicle Loading and Cost sections with formulas entered in the first row

To complete the empty cells in rows 17 through 20, you can copy the formulas from cells G16 through N16 to the rest of the rows. Click and drag to select cells G16 through N16, then right-click and select Copy from the menu (see Figure D-11).

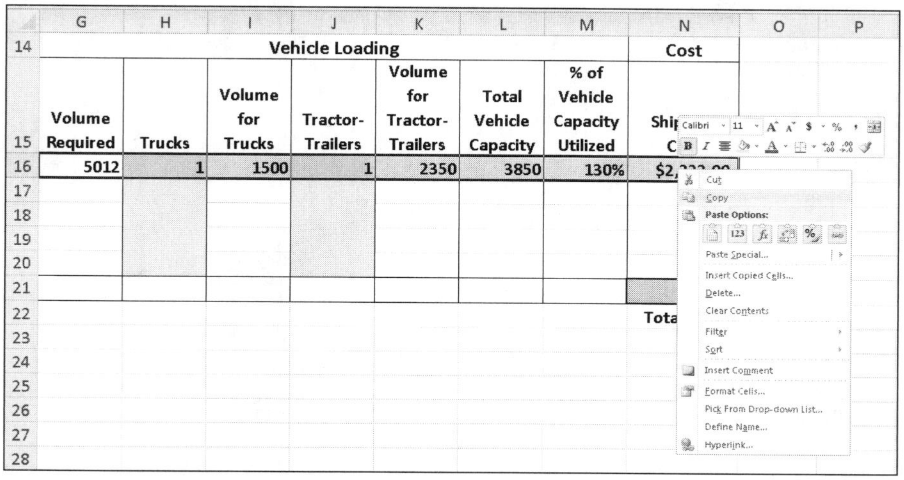

FIGURE D-11 Copying formulas

Next, select cells G17 through N20, which are in the four rows beneath row 16. Either press Enter or click Paste in the Clipboard group. The formulas from row 16 should be copied to the rest of the destination cities (see Figure D-12).

				Vehicle Loading				Cost
Volume Required	Trucks	Volume for Trucks	Tractor-Trailers	Volume for Tractor-Trailers	Total Vehicle Capacity	% of Vehicle Capacity Utilized	Shipping Cost
5012	1	1500	1	2350	3850	130%	$2,323.00
3513	1	1500	1	2350	3850	91%	$874.00
2776	1	1500	1	2350	3850	72%	$2,300.00
6630	1	1500	1	2350	3850	172%	$1,242.00
7785	1	1500	1	2350	3850	202%	$4,163.00
							Total Cost

FIGURE D-12 Formulas from row 16 successfully copied to rows 17 through 20

You have one row of formulas to complete: the Totals row. You will use the AutoSum function to sum up one column, and then copy the formula to the rest of the columns *except* cell M21. This cell is not actually a total, but an overall capacity utilization rate.

To enter the sum of cells G16 through G20 in cell G21, select cells G16 through G21, then click AutoSum in the Editing group on the right edge of the Ribbon (see Figure D-13).

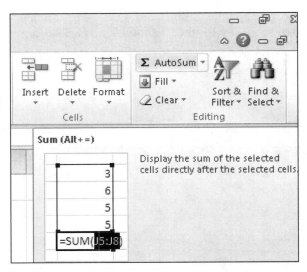

FIGURE D-13 AutoSum button in the Editing group

Cell G21 should now contain the formula =SUM(G16:G20), and the displayed answer should be 25716. Now you can copy cell G21 to cells H21, I21, J21, K21, L21, and N21. When you have completed this section of the table, it should have the values shown in Figure D-14.

	G	H	I	J	K	L	M	N
14				Vehicle Loading				Cost
15	Volume Required	Trucks	Volume for Trucks	Tractor-Trailers	Volume for Tractor-Trailers	Total Vehicle Capacity	% of Vehicle Capacity Utilized	Shipping Cost
16	5012	1	1500	1	2350	3850	130%	$2,323.00
17	3513	1	1500	1	2350	3850	91%	$874.00
18	2776	1	1500	1	2350	3850	72%	$2,300.00
19	6630	1	1500	1	2350	3850	172%	$1,242.00
20	7785	1	1500	1	2350	3850	202%	$4,163.00
21	25716	5	7500	5	11750	19250		$10,902.00
22								Total Cost

FIGURE D-14 Most totals cells completed

The last formula to enter is for cell M21. This is not a total, but an overall percentage of Vehicle Capacity Utilized for all the vehicles used. This calculation uses the same formula as the cell above, so you can simply copy cell M20 to cell M21. The formula for this cell is =G21/L21, which is Volume Required divided by Total Vehicle Capacity, expressed as a percentage. Your completed spreadsheet should look like Figure D-15.

	A	B	C	D	E	F	G	H	I	J	K	L	M	N
12														
13		Calculations and Results Section:												
14		Distance/Demand Table			Store Demand					Vehicle Loading				Cost
15		Distance Table (from Memphis Plant)	Miles	EB	CT	TM	Volume Required	Trucks	Volume for Trucks	Tractor-Trailers	Volume for Tractor-Trailers	Total Vehicle Capacity	% of Vehicle Capacity Utilized	Shipping Cost
16		Philadelphia Store	1010	140	96	86	5012	1	1500	1	2350	3850	130%	$2,323.00
17		Atlanta Store	380	76	81	63	3513	1	1500	1	2350	3850	91%	$874.00
18		Miami Store	1000	56	64	52	2776	1	1500	1	2350	3850	72%	$2,300.00
19		Chicago Store	540	115	130	150	6630	1	1500	1	2350	3850	172%	$1,242.00
20		Los Angeles Store	1810	150	135	180	7785	1	1500	1	2350	3850	202%	$4,163.00
21						Totals:	25716	5	7500	5	11750	19250	134%	$10,902.00
22		Fill Legend:			Changing Cells									Total Cost
23					Optimization Cell									

FIGURE D-15 Completed Calculations and Results section

Working the Model Manually

Now that you have a working model, you could manipulate the number of trucks and tractor-trailers manually to obtain a solution to the shipping problem. You would need to observe the following rules (or constraints):

1. Assign enough Total Vehicle Capacity to meet the Volume Required for each destination. (In other words, you cannot exceed 100% of Vehicle Capacity Utilized.)
2. The total number of trucks and tractor-trailers you assign cannot exceed the number available in your fleet.

Try to assign your trucks and tractor-trailers to meet your shipping requirements, and note the total shipping costs—you may get lucky and come up with an optimal solution. The tractor-trailers are more cost efficient than the trucks, but the problem is complicated by the fact that you want to achieve the best capacity utilization as well. In some instances, the trucks may be a better fit. Figure D-16 shows a sample solution determined from working the problem manually.

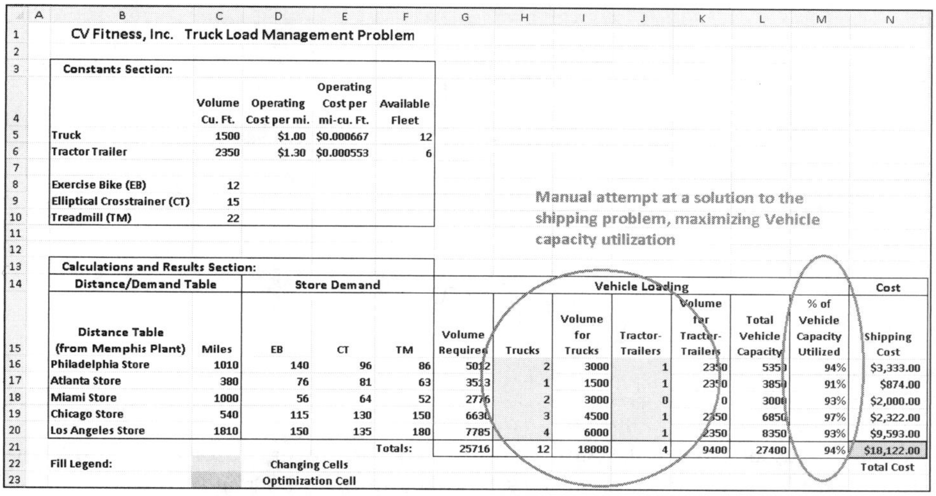

FIGURE D-16   Manual attempt to solve the vehicle loading problem optimally

This probably looks like a good solution—after all, you have not violated any of your constraints, and you have a 94% average vehicle capacity utilization. But is it the most cost-effective solution for your company? This is where Solver comes in.

Setting up Solver Using the Solver Parameters Window

To access the Solver pane, click the Data tab on the Ribbon, then click Solver in the Analysis group on the far right side of the Ribbon. The Solver Parameters window appears (see Figure D-17).

> **NOTE**
>
> Solver in Excel 2010 has changed significantly from earlier versions of Excel. It allows three different calculation methods, and allows you to specify an amount of time and number of iterations to perform before Excel ends the calculation. Refer to Microsoft Help for more information on Solver's new capabilities.

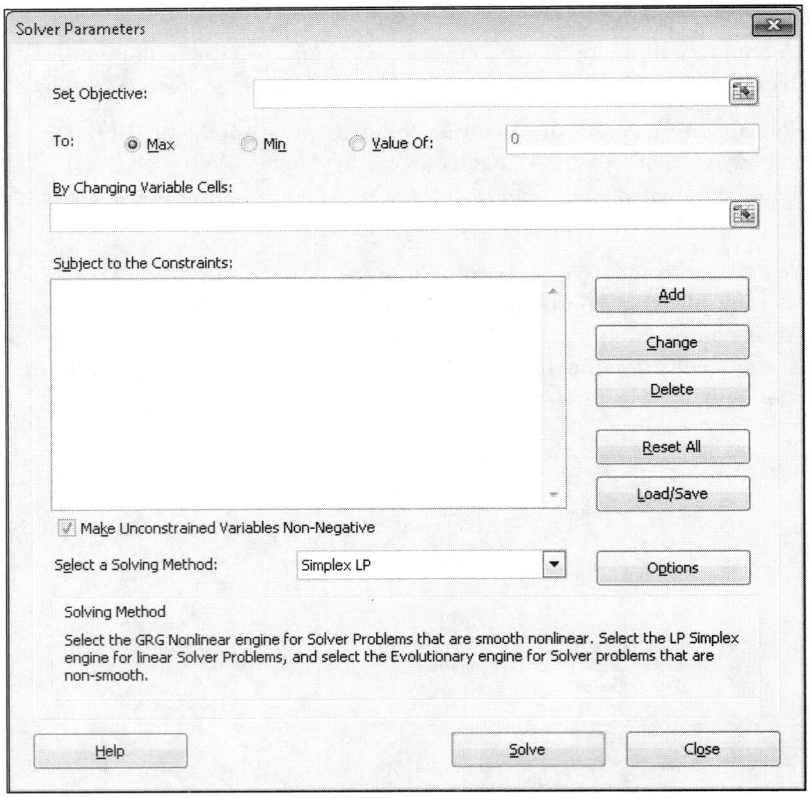

FIGURE D-17   Solver Parameters window

The Solver Parameters window in Excel 2010 looks intimidating at first. However, to solve linear optimization problems, you only have to satisfy three sets of conditions by filling in the following fields:

- Set Objective—Specify the Optimization Cell.
- By Changing Variable Cells—Specify the Changing Cells in your worksheet.
- Subject to the Constraints—Define all of the conditions and limitations that must be met when seeking the optimal solution.

The following sections explain these fields in detail. You may also need to click the Options button and select one or more options for solving the problem. Most of the cases in this book are linear problems, so you can set the solving method to Simplex LP, as shown in Figure D-17. If this method does not work in later cases, you can select the GRG Nonlinear or Evolutionary method to try to solve the problem. Note that solving methods are only available in Excel 2010.

Optimization Cell and Changing Cells

To use Solver successfully, you must first specify the cell you want to optimize—in this case, the total shipping cost, or cell N21. To fill the Set Objective field, click the button at the right edge of the field, and then click cell N21 in the spreadsheet. You could also type the cell address in the window, but selecting the cell in the spreadsheet reduces your chance of entering the wrong cell address. Next, specify whether you want Solver to seek the maximum or minimum value for cell N21. Because you want to minimize the total shipping cost, click the radio button next to Min.

Next, tell Solver which cell values it will change to determine the optimal solution. Use the By Changing Variable Cells field to specify the range of cells that you want Solver to manipulate. Again, click the button at the right edge of the field, select the cells that contain the numbers of trucks (H16 to H20), and then hold down the Ctrl key and select the cells that contain the numbers of tractor-trailers (J16 to J20). If you used a fill color for the Changing Cells, they will be easy to find and select. The Solver Parameters window should look like Figure D-18.

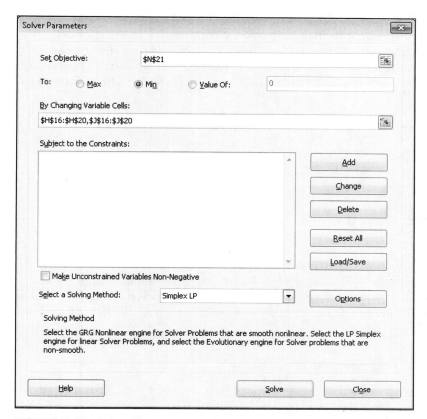

FIGURE D-18   Solver Parameters window with the Objective cell and Changing Cells entered

Note that Solver has added absolute cell references (the $ signs before the column and row designators) for the cells you have specified. Solver will also add these references to the constraints you define. Solver adds the references to preserve the links to the cells in case you revise the worksheet in the future. In fact, you will make changes to the worksheet later in the tutorial.

Defining and Entering Constraints

For Solver to successfully determine the optimum solution for the shipping problem, you need to specify what constraints or rules it must observe to calculate the solution. Without constraints, Solver theoretically might calculate that the best solution is not to ship anything, resulting in a cost of zero. Furthermore, if you failed to define variables as positive numbers, Solver would select "negative trucks" to maximize "negative costs." Finally, the trucks are indivisible units—you cannot assign a fraction of a truck for a fraction of the cost, so you must define your Changing Cells as integers to satisfy this constraint.

Aside from the preceding logical constraints, you have operational constraints as well. You cannot assign more vehicles than you have in your fleet, and the vehicles you assign must have at least as much total capacity as your shipping volume.

Before entering the constraints in the Solver Parameters window, it is generally a good idea to list them in plain terms. You must enter the following constraints for this model:

- All trucks and tractor-trailers in the Changing Cells must be integers greater than or equal to zero.
- The sums of trucks and tractor-trailers assigned (cells H21 and J21) must be less than or equal to the available trucks and tractor-trailers (cells F5 and F6, respectively).
- The Total Vehicle Capacity for the vehicles assigned to each store (cells L16 to L20) must be greater than or equal to the Volume Required to be shipped to each store (cells G16 to G20, respectively).

You are ready to enter the constraints as equations or inequalities in the Add Constraint dialog box. To begin, click the Add button in the Solver Parameters window. In the dialog box that appears (see Figure D-19), click the button at the right edge of the Cell Reference box, select cells H16 to H20, and then click the button again. Next, click the drop-down menu in the middle field and select >=. Then go to the Constraint field and type 0. Finally, click Add; otherwise, the constraint you defined will not be added to the list defined in the Solver Parameters window.

FIGURE D-19   Add Constraint dialog box

You can continue to add constraints in the Add Constraint dialog box. For this example, enter the constraints shown in the completed Solver Parameters window in Figure D-20. When you finish, click Add to save the last constraint, then click Cancel in the Add Constraint dialog box to return to the Solver Parameters window.

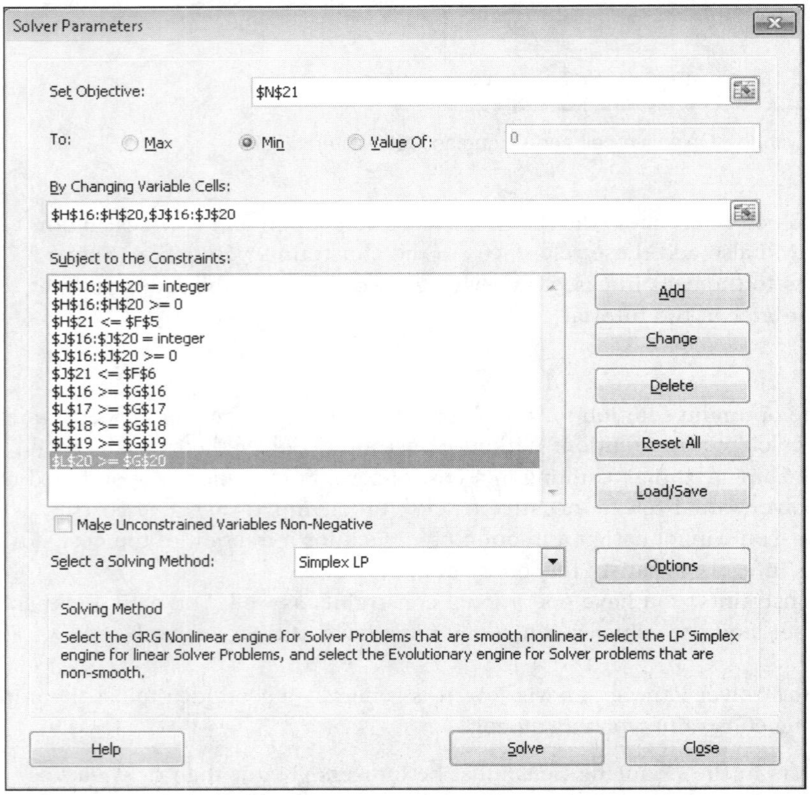

FIGURE D-20   Completed Solver Parameters window

If you have difficulty reading the constraints listed in Figure D-20, use the following list instead:

- $H$16:$H$20 = integer
- $H$16:$H$20 >= 0
- $H21 <= $F$5

- $J$16:$J$20 = integer
- $J$16:$J$20 >= 0
- $J21 <= $F$6
- $L$16 >= $G$16
- $L$17 >= $G$17
- $L$18 >= $G$18
- $L$19 >= $G$19
- $L$20 >= $G$20

You should also click the Options button in the Solver Parameters window and check the Options dialog box shown in Figure D-21. You can use this dialog box to set the maximum amount of time and iterations you want Solver to run before stopping. Leave both options at 100 for now, but remember that Solver may need more time and iterations for more complex problems. To get the best solution, you should set the Integer Optimality (%) to zero. Click OK to close the dialog box.

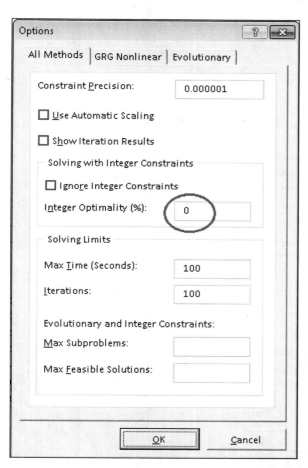

FIGURE D-21   Solver Options dialog box with Integer Optimality set to 0

You are ready to run Solver to find the optimal solution. Click Solve at the bottom of the Solver Parameters window. Solver might require only a few seconds or more than a minute to run all the possible iterations—the status bar at the bottom of the Excel window displays iterations and possible solutions continuously until Solver finds an optimal solution or runs out of time (see Figure D-22).

33					
34					

◄ ► ► ► Sheet1　Sheet2　Sheet3

Incumbent: $20,832.00　Subproblem: 168　Trial Solution: 0　Objective Cell: $20,832.00

FIGURE D-22　Excel status bar showing Solver running through possible solutions

A new dialog box will appear eventually, indicating that Solver has found an optimal solution to the problem (see Figure D-23). The portion of the spreadsheet that displays the assigned vehicles and shipping cost is visible below the Solver Results dialog box. Solver has assigned nine of the 12 trucks and all six tractor-trailers, for a total shipping cost of $17,398. The earlier manual attempt to solve the problem (see Figure D-16) assigned all 12 trucks and four tractor-trailers, for a total shipping cost of $18,122. Using Solver saved your company $724.

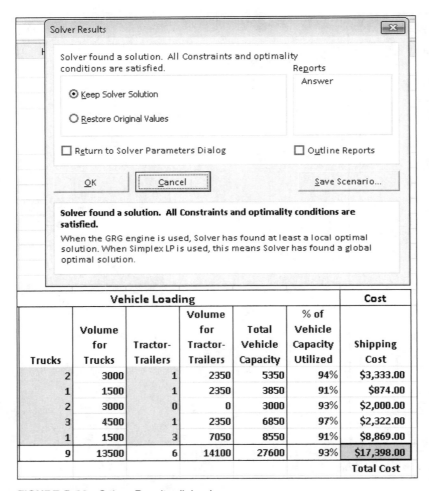

	Vehicle Loading					Cost
Trucks	Volume for Trucks	Tractor-Trailers	Volume for Tractor-Trailers	Total Vehicle Capacity	% of Vehicle Capacity Utilized	Shipping Cost
2	3000	1	2350	5350	94%	$3,333.00
1	1500	1	2350	3850	91%	$874.00
2	3000	0	0	3000	93%	$2,000.00
3	4500	1	2350	6850	97%	$2,322.00
1	1500	3	7050	8550	91%	$8,869.00
9	13500	6	14100	27600	93%	$17,398.00
						Total Cost

FIGURE D-23　Solver Results dialog box

If the Solver Results dialog box does not report an optimal solution to the problem, it will report that the problem could not be solved given the Changing Cells and constraints you specified. For instance, if you had not had enough vehicles in your fleet to carry the required shipping volume to all the destinations, the Solver Results dialog box might have looked like Figure D-24. In the figure, your vehicle fleet was reduced to 10 trucks and five tractor-trailers, so Solver could not find a solution that satisfied the shipping volume constraints.

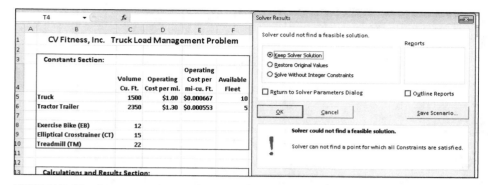

FIGURE D-24   Solver could not find a feasible solution with a reduced vehicle fleet

Fortunately, Solver did find an optimal solution. To update the spreadsheet with the new optimal values for the Changing Cells and the Optimization Cell, click OK in the Solver Results dialog box. You can also print an Answer Report by clicking the Answer option in the Solver Results dialog box (see Figure D-25) and then clicking OK.

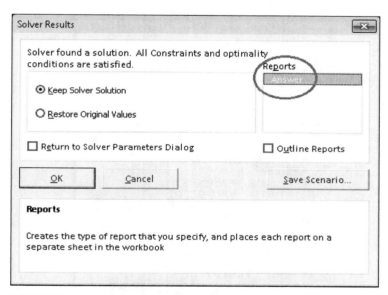

FIGURE D-25   Printing an Answer Report

Excel will create a report in a separate sheet called Answer Report 1. The Answer Report is shown in Figures D-26 and D-27.

	A	B	C	D	E	F
1	Microsoft Excel 14.0 Answer Report					
2	Worksheet: [CV Fitness Trucking Problem--before Solver--Five Stores.xlsx]Sheet1					
3	Report Created: 11/5/2010 6:17:03 PM					
4	Result: Solver found a solution. All Constraints and optimality conditions are satisfied.					
5	Solver Engine					
6	Engine: Simplex LP					
7	Solution Time: 2.714 Seconds.					
8	Iterations: 3 Subproblems: 1416					
9	Solver Options					
10	Max Time 100 sec, Iterations 100, Precision 0.000001					
11	Max Subproblems Unlimited, Max Integer Sols Unlimited, Integer Tolerance 0%					
12						
13	Objective Cell (Min)					
14	**Cell**		**Name**	**Original Value**	**Final Value**	
15	$N$21		Totals: Shipping Cost	$18,122.00	$17,398.00	
16						
17	Variable Cells					
18	**Cell**		**Name**	**Original Value**	**Final Value**	**Integer**
19	$H$16		Philadelphia Store Trucks	2	2	Integer
20	$H$17		Atlanta Store Trucks	1	1	Integer
21	$H$18		Miami Store Trucks	2	2	Integer
22	$H$19		Chicago Store Trucks	3	3	Integer
23	$H$20		Los Angeles Store Trucks	4	1	Integer
24	$J$16		Philadelphia Store Tractor-Trailers	1	1	Integer
25	$J$17		Atlanta Store Tractor-Trailers	1	1	Integer
26	$J$18		Miami Store Tractor-Trailers	0	0	Integer
27	$J$19		Chicago Store Tractor-Trailers	1	1	Integer
28	$J$20		Los Angeles Store Tractor-Trailers	1	3	Integer
29						

FIGURE D-26   Top portion of the Answer Report

	A	B	C	D	E	F	G
30	Constraints						
31	**Cell**		**Name**	**Cell Value**	**Formula**	**Status**	**Slack**
32	$L$19		Chicago Store Total Vehicle Capacity	6850	$L$19>=$G$19	Not Binding	220
33	$L$17		Atlanta Store Total Vehicle Capacity	3850	$L$17>=$G$17	Not Binding	337
34	$L$16		Philadelphia Store Total Vehicle Capacity	5350	$L$16>=$G$16	Not Binding	338
35	$J$21		Totals: Tractor-Trailers	6	$J$21<=$F$6	Binding	0
36	$H$21		Totals: Trucks	9	$H$21<=$F$5	Not Binding	3
37	$L$18		Miami Store Total Vehicle Capacity	3000	$L$18>=$G$18	Not Binding	224
38	$L$20		Los Angeles Store Total Vehicle Capacity	8550	$L$20>=$G$20	Not Binding	765
39	$H$16		Philadelphia Store Trucks	2	$H$16>=0	Binding	0
40	$H$17		Atlanta Store Trucks	1	$H$17>=0	Binding	0
41	$H$18		Miami Store Trucks	2	$H$18>=0	Binding	0
42	$H$19		Chicago Store Trucks	3	$H$19>=0	Binding	0
43	$H$20		Los Angeles Store Trucks	1	$H$20>=0	Binding	0
44	$J$16		Philadelphia Store Tractor-Trailers	1	$J$16>=0	Binding	0
45	$J$17		Atlanta Store Tractor-Trailers	1	$J$17>=0	Binding	0
46	$J$18		Miami Store Tractor-Trailers	0	$J$18>=0	Binding	0
47	$J$19		Chicago Store Tractor-Trailers	1	$J$19>=0	Binding	0
48	$J$20		Los Angeles Store Tractor-Trailers	3	$J$20>=0	Binding	0
49	$H$16:$H$20=Integer						
50	$J$16:$J$20=Integer						

I◄ ◄ ► ►I   **Answer Report 1**   Truck Loading Problem

FIGURE D-27 . Bottom portion of the Answer Report—note the new tab created by Solver

The Answer Report gives you a wealth of information about the solution. The top portion displays the original and final values of the Objective cell. The second part of the report displays the original and final values of the Changing Cells. The last part of the report lists the constraints. Binding constraints are those that reached their maximum or minimum value; nonbinding constraints did not.

Perhaps a savings of $724 does not seem significant—however, this problem does not have a specified time frame. The example probably represents one week of shipments for CV Fitness. The store demands will change from week to week, but you could use Solver each time to optimize the truck assignments. In a 50-week business year, the savings in shipping costs from using Solver could be well over $30,000!

Go to the File tab to print the worksheets you created. Save the Excel file as **CV Fitness Trucking Problem.xlsx**, then select the Save As command in the File tab to create a new file called **CV Fitness Trucking Problem 2.xlsx**. You will use the new file in the next section.

EXTENDING THE EXAMPLE

Like all successful companies, CV Fitness looks for ways to grow its business and optimize its costs. Your management team is considering two changes:

- Opening two new stores and expanding the vehicle fleet if necessary
- Improving product design and packaging to reduce the shipping volume of the treadmill from 22 cubic feet to 17 cubic feet

You have been asked to modify your model to see the new requirements for each change separately. The two new stores are in Denver and Phoenix, and they are 1040 and 1470 miles from the Memphis plant, respectively. If necessary, open the CV Fitness Trucking Problem 2.xlsx file, then right-click the 21 at the left worksheet border to select row 21. Click Insert to enter a new row between rows 20 and 21. Repeat the steps to insert a second new row. Your spreadsheet should look like Figure D-28. Do not worry about the borders for now—you can fix them later.

	Distance Table (from Memphis Plant)	Miles	EB	CT	TM	Volume Required	Trucks	Volume for Trucks	Tractor-Trailers	Volume for Tractor-Trailers	Total Vehicle Capacity	% of Vehicle Capacity Utilized	Shipping Cost
Philadelphia Store	1010	140	96	86	5012	2	3000	1	2350	5350	94%	$3,333.00
Atlanta Store	380	76	81	63	3513	1	1500	1	2350	3850	91%	$874.00
Miami Store	1000	56	64	52	2776	2	3000	0	0	3000	93%	$2,000.00
Chicago Store	540	115	130	150	6630	3	4500	1	2350	6850	97%	$2,322.00
Los Angeles Store	1810	150	135	180	7785	1	1500	3	7050	8550	91%	$8,869.00
					Totals:	25716	9	13500	6	14100	27600	93%	$17,398.00

FIGURE D-28   Distance/Demand table with two blank rows inserted for the new stores

Enter the two new stores in cells B21 and B22, enter their distances in cells C21 and C22, and enter the Store Demands in cells D21 through F22, as shown in Figure D-29.

	Distance Table (from Memphis Plant)	Miles	EB	CT	TM
	Philadelphia Store	1010	140	96	86
	Atlanta Store	380	76	81	63
	Miami Store	1000	56	64	52
	Chicago Store	540	115	130	150
	Los Angeles Store	1810	150	135	180
	Denver Store	1040	74	67	43
	Phoenix Store	1470	41	28	37
				Totals:	

FIGURE D-29   Distance/Demand table with new store locations and demands entered

Next, copy the formulas from cells G20 to N20 to the two new rows in the Vehicle Loading and Cost sections of the table. Select cells G20 to N20, right-click, and click Copy on the menu. Then select cells G21 to N22 and click Paste in the Clipboard group. Your table should look like Figure D-30.

	G	H	I	J	K	L	M	N
13								
14			Vehicle Loading					Cost
15	Volume Required	Trucks	Volume for Trucks	Tractor-Trailers	Volume for Tractor-Trailers	Total Vehicle Capacity	% of Vehicle Capacity Utilized	Shipping Cost
16	5012	2	3000	1	2350	5350	94%	$3,333.00
17	3513	1	1500	1	2350	3850	91%	$874.00
18	2776	2	3000	0	0	3000	93%	$2,000.00
19	6630	3	4500	1	2350	6850	97%	$2,322.00
20	7785	1	1500	3	7050	8550	91%	$8,869.00
21	2839	1	1500	3	7050	8550	33%	$5,096.00
22	1726	1	1500	3	7050	8550	20%	$7,203.00
23	30281	9	13500	6	14100	27600	110%	$17,398.00
24								Total Cost

FIGURE D-30 Formulas from row 20 copied into rows 21 and 22

Note that most cells in the Totals row have not changed—their formulas need to be updated to include rows 21 and 22. To quickly check which cells you need to update, display the formulas in the Totals row. Hold down the Ctrl key and press the ~ key; on most keyboards, this key is near the upper-left corner next to the "1" key. The Vehicle Loading and Cost sections now display formulas in the cells (see Figure D-31).

	G	H	I	J	K	L	M	N
14			Vehicle Loading					Cost
15	Volume Required	Trucks	Volume for Trucks	Tractor-Trailers	Volume for Tractor-Trailers	Total Vehicle Capacity	% of Vehicle Capacity Utilized	Shipping Cost
16	=D16*$C$8+E16*$C$9	2	=H16*$C$5	1	=J16*$C$6	=I16+K16	=G16/L16	=H16*C16*$D$5+J16*C16*D
17	=D17*$C$8+E17*$C$9	1	=H17*$C$5	1	=J17*$C$6	=I17+K17	=G17/L17	=H17*C17*$D$5+J17*C17*D
18	=D18*$C$8+E18*$C$9	2	=H18*$C$5	0	=J18*$C$6	=I18+K18	=G18/L18	=H18*C18*$D$5+J18*C18*D
19	=D19*$C$8+E19*$C$9	3	=H19*$C$5	1	=J19*$C$6	=I19+K19	=G19/L19	=H19*C19*$D$5+J19*C19*D
20	=D20*$C$8+E20*$C$9	1	=H20*$C$5	3	=J20*$C$6	=I20+K20	=G20/L20	=H20*C20*$D$5+J20*C20*D
21	=D21*$C$8+E21*$C$9	1	=H21*$C$5	3	=J21*$C$6	=I21+K21	=G21/L21	=H21*C21*$D$5+J21*C21*D
22	=D22*$C$8+E22*$C$9	1	=H22*$C$5	3	=J22*$C$6	=I22+K22	=G22/L22	=H22*C22*$D$5+J22*C22*D
23	=SUM(G16:G22)	=SUM(H16:H20)	=SUM(I16:I20)	=SUM(J16:J20)	=SUM(K16:K20)	=SUM(L16:L20)	=G23/L23	=SUM(N16:N20)
24								Total Cost

FIGURE D-31 Vehicle Loading and Cost sections with formulas displayed in the cells

You must update any Totals cells that do not include the contents of rows 21 and 22. For example, you need to update the Totals cells H23 through L23 and cell N23. Cell M23 is not really a total; it is a cumulative ratio formula, so you do not need to update the cell. Use the following formulas to revise the Totals cells:

- Cell H23: =SUM(H16:H22)
- Cell I23: =SUM(I16:I22)
- Cell J23: =SUM(J16:J22)
- Cell K23: =SUM(K16:K22)
- Cell L23: =SUM(L16:L22)
- Cell N23: =SUM(N16:N22)

The updated sections should look like Figure D-32.

	G	H	I	J	K	L	M	N
14	Vehicle Loading							Cost
15	Volume Required	Trucks	Volume for Trucks	Tractor-Trailers	Volume for Tractor-Trailers	Total Vehicle Capacity	% of Vehicle Capacity Utilized	Shipping Cost
16	5012	2	3000	1	2350	5350	94%	$3,333.00
17	3513	1	1500	1	2350	3850	91%	$874.00
18	2776	2	3000	0	0	3000	93%	$2,000.00
19	6630	3	4500	1	2350	6850	97%	$2,322.00
20	7785	1	1500	3	7050	8550	91%	$8,869.00
21	2839	1	1500	3	7050	8550	33%	$5,096.00
22	1726	1	1500	3	7050	8550	20%	$7,203.00
23	30281	11	16500	12	28200	44700	68%	$29,697.00
24								Total Cost

FIGURE D-32 Vehicle Loading and Cost sections with the formulas updated

You are ready to use Solver to determine the optimal vehicle assignment. Click Solver in the Analysis group of the Data tab. You should notice immediately that you must change the Changing Cells to include the two new stores; you must also change some of the constraints and add others. Solver has already updated the Objective cell from $N$21 to $N$23 and has updated the $H$23<=$F$5 and $J$23<=$F$6 constraints for vehicle fleet size. To update the Changing Cells, click the button to the right of the By Changing Variable Cells field and select the cells again, or edit the formula in the window by changing cell address $H$20 to $H$22 and cell address $J$20 to $J$22.

To change a constraint, select the one you want to change, and then click Change (see Figure D-33).

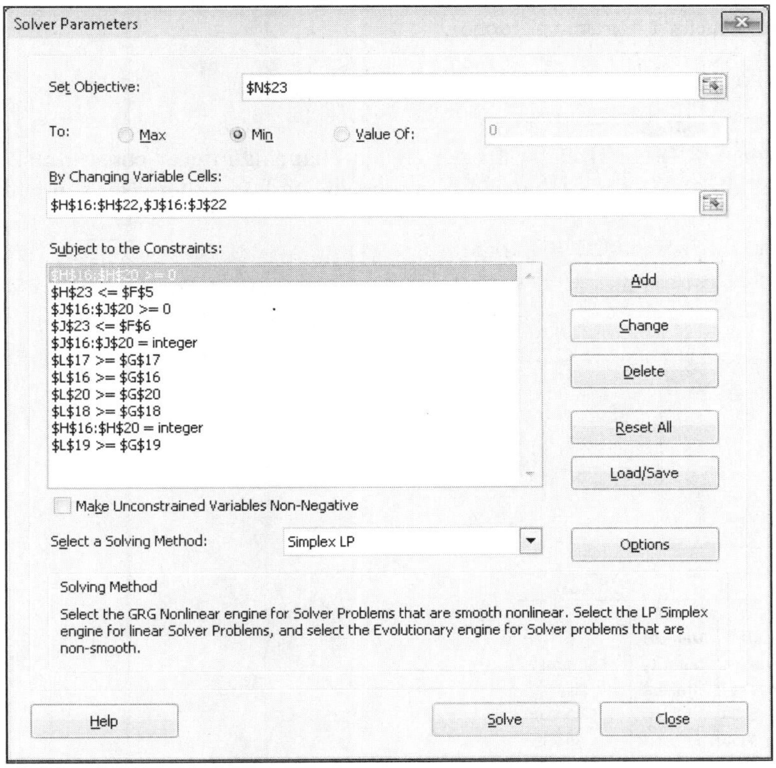

FIGURE D-33 Selecting a constraint to change

When you click Change, the Change Constraint dialog box appears. Click the Cell Reference button; the selected cells will appear on the spreadsheet with a moving marquee around them (see Figure D-34). Highlight the new group of cells; when the new range appears in the Cell Reference field, click OK. The Solver Parameters window appears with the constraint changed.

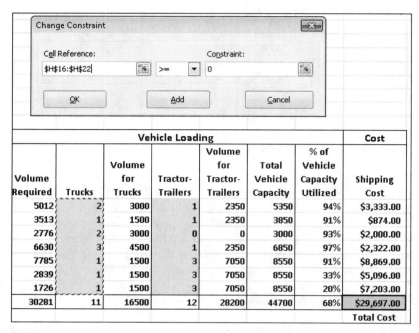

FIGURE D-34   Adding cells H21 and H22 to the Trucks constraint cell range

You also need to update or add the following constraints:

- Update $J$16:$J$20 >=0 to $J$16:$J$22 >=0.
- Update $H$16:$H$20 = integer to $H$16:$H$22 = integer. When changing integer constraints, you must click "int" in the middle field of the Change Constraint dialog box; otherwise, you will receive an error message.
- Update $J$16:$J$20 = integer to $J$16:$J$22 = integer.
- Add constraint $L$21 >= $G$21 (see Figure D-35).
- Add constraint $L$22 >= $G$22.

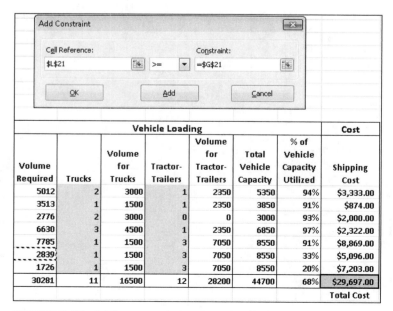

FIGURE D-35   Adding a constraint using the Add Constraint dialog box

You are ready to solve the shipping problem to include the new stores in Denver and Phoenix. Figure D-36 shows the updated Solver Parameters window.

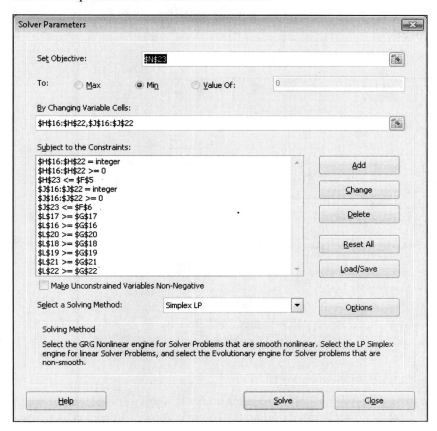

FIGURE D-36   Solver parameters updated for shipping to seven stores

Before you run Solver again, you might want to attempt to assign the trucks manually, because your fleet may not be large enough to handle two more stores. In this case, you will quickly realize that the vehicle fleet is at least one truck or tractor-trailer short of the minimum required to ship the needed volume. You can confirm this by running Solver (see Figure D-37).

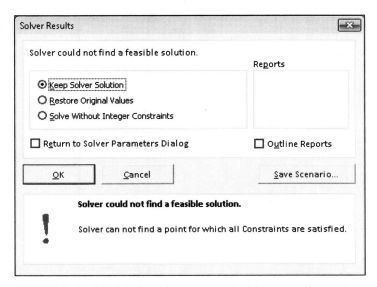

FIGURE D-37   Vehicle fleet does not meet minimum requirements

The Solver Results window confirms that your truck fleet is too small, so add another truck to your fleet and run Solver again. As you add more stores and vehicles to make the problem more complex, Solver will take longer to run, especially on older computers. You may have to wait a minute or more for Solver to finish its iterations and find an answer (see Figure D-38). In this example, Solver recommends that you use 13 trucks and six tractor-trailers.

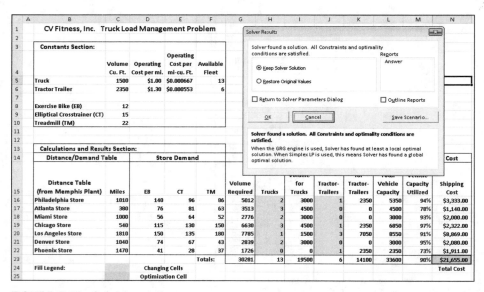

FIGURE D-38   Solver's solution

Select Answer in the Reports list to add an Answer Report to the workbook, and then click OK. You can keep or delete the old Answer Report 1 tab from the earlier workbook. The new Answer Report is in a new worksheet named Answer Report 2.

You can meet the shipping requirements by adding one more truck, but is it really the most cost-effective solution? What if you add a tractor-trailer instead? Set the number of trucks back to 12, and add a tractor-trailer by entering 7 instead of 6 in cell F6. Run Solver again.

This time Solver finds a less expensive solution, as shown in Figures D-39 and D-40. At first it does not make sense—how can adding a more expensive vehicle (a tractor-trailer) reduce the overall expense? In fact, the additional tractor-trailer has replaced two trucks. With seven tractor-trailers, you only need 11 trucks instead of the original 13.

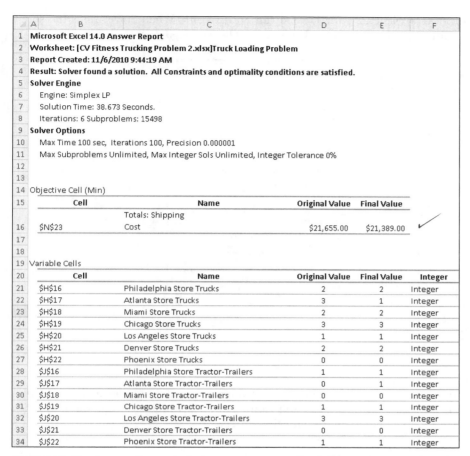

	A	B	C	D	E	F
1	Microsoft Excel 14.0 Answer Report					
2	Worksheet: [CV Fitness Trucking Problem 2.xlsx]Truck Loading Problem					
3	Report Created: 11/6/2010 9:44:19 AM					
4	Result: Solver found a solution. All Constraints and optimality conditions are satisfied.					
5	Solver Engine					
6	Engine: Simplex LP					
7	Solution Time: 38.673 Seconds.					
8	Iterations: 6 Subproblems: 15498					
9	Solver Options					
10	Max Time 100 sec, Iterations 100, Precision 0.000001					
11	Max Subproblems Unlimited, Max Integer Sols Unlimited, Integer Tolerance 0%					
12						
13						
14	Objective Cell (Min)					
15	Cell	Name		Original Value	Final Value	
		Totals: Shipping				
16	$N$23	Cost		$21,655.00	$21,389.00	
17						
18						
19	Variable Cells					
20	Cell	Name		Original Value	Final Value	Integer
21	$H$16	Philadelphia Store Trucks		2	2	Integer
22	$H$17	Atlanta Store Trucks		3	1	Integer
23	$H$18	Miami Store Trucks		2	2	Integer
24	$H$19	Chicago Store Trucks		3	3	Integer
25	$H$20	Los Angeles Store Trucks		1	1	Integer
26	$H$21	Denver Store Trucks		2	2	Integer
27	$H$22	Phoenix Store Trucks		0	0	Integer
28	$J$16	Philadelphia Store Tractor-Trailers		1	1	Integer
29	$J$17	Atlanta Store Tractor-Trailers		0	1	Integer
30	$J$18	Miami Store Tractor-Trailers		0	0	Integer
31	$J$19	Chicago Store Tractor-Trailers		1	1	Integer
32	$J$20	Los Angeles Store Tractor-Trailers		3	3	Integer
33	$J$21	Denver Store Tractor-Trailers		0	0	Integer
34	$J$22	Phoenix Store Tractor-Trailers		1	1	Integer

FIGURE D-39   Answer Report 3 displays a more cost-effective solution

Vehicle Loading							Cost
Volume Required	Trucks	Volume for Trucks	Tractor-Trailers	Volume for Tractor-Trailers	Total Vehicle Capacity	% of Vehicle Capacity Utilized	Shipping Cost
5012	2	3000	1	2350	5350	94%	$3,333.00
3513	1	1500	1	2350	3850	91%	$874.00
2776	2	3000	0	0	3000	93%	$2,000.00
6630	3	4500	1	2350	6850	97%	$2,322.00
7785	1	1500	3	7050	8550	91%	$8,869.00
2839	2	3000	0	0	3000	95%	$2,080.00
1726	0	0	1	2350	2350	73%	$1,911.00
30281	11	16500	7	16450	32950	92%	$21,389.00
							Total Cost

FIGURE D-40   Seven tractor-trailers and 11 trucks are the optimal mix

You have a solution for the expansion to seven stores. Save your workbook, and then create a new workbook using the Save As command. Name the new workbook **CV Fitness Trucking Problem 3.xlsx**.

Next, evaluate the potential cost savings if the company redesigns its treadmill product and packaging to reduce the shipping volume from 22 cubic feet to 17 cubic feet. Your engineers report that the redesign will cost approximately $10,000. If you can save at least $500 per shipment, the project will pay for itself in less than six months (20 weekly shipments).

Go to cell C10 on the worksheet, replace 22 with 17, and run Solver again. When Solver finds the solution, select Answer to create another Answer Report, and then click OK. See Figure D-41.

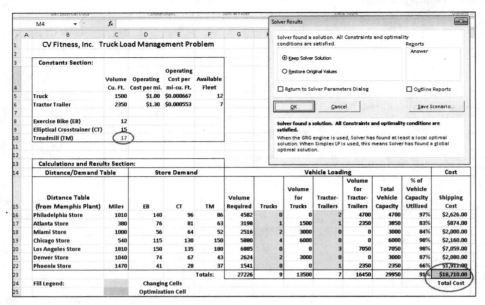

FIGURE D-41 Solver solution with redesigned treadmill and packaging

Check the Answer Report to see the cost difference between shipping the old treadmills and the redesigned models (see Figure D-42). The cost savings for one shipment is $2,679, which is more than five times the minimum savings you needed. You should go ahead with the project.

	A	B	C	D	E	F
1	Microsoft Excel 14.0 Answer Report					
2	Worksheet: [CV Fitness Trucking Problem 3.xlsx]Truck Loading Problem					
3	Report Created: 11/6/2010 10:14:10 AM					
4	Result: Solver found a solution. All Constraints and optimality conditions are satisfied.					
5	Solver Engine					
6	Engine: Simplex LP					
7	Solution Time: 25.21 Seconds.					
8	Iterations: 8 Subproblems: 10238					
9	Solver Options					
10	Max Time 100 sec, Iterations 100, Precision 0.000001					
11	Max Subproblems Unlimited, Max Integer Sols Unlimited, Integer Tolerance 0%					
12						
13	Objective Cell (Min)					
14	**Cell**	**Name**		**Original Value**	**Final Value**	
15	$N$23	Totals: Shipping Cost		$21,389.00	$18,710.00	
16						
17	Variable Cells					
18	**Cell**	**Name**		**Original Value**	**Final Value**	**Integer**
19	$H$16	Philadelphia Store Trucks		2	0 Integer	
20	$H$17	Atlanta Store Trucks		1	1 Integer	
21	$H$18	Miami Store Trucks		2	2 Integer	
22	$H$19	Chicago Store Trucks		3	4 Integer	
23	$H$20	Los Angeles Store Trucks		1	0 Integer	
24	$H$21	Denver Store Trucks		2	2 Integer	
25	$H$22	Phoenix Store Trucks		0	0 Integer	

FIGURE D-42 Answer Report for the treadmill redesign

When you finish examining the Answer Report, save your file and then close it. To close the workbook, click the File tab and then click Close (see Figure D-43).

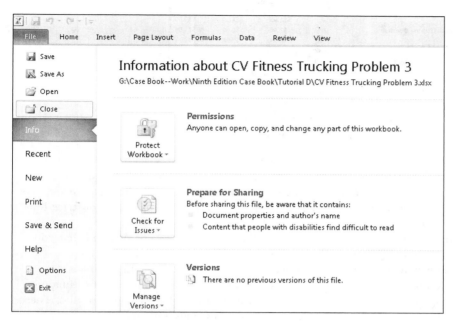

FIGURE D-43   Closing the Excel workbook

USING SOLVER ON A NEW PROBLEM

A common problem in manufacturing businesses is deciding on a product mix for different items in the same product family. Sensuous Scents Inc. makes a premium collection of perfume, cologne, and body spray for sale in large department stores and boutiques. The primary ingredient is ambergris, a valuable digestive excretion from whales that is harvested without harming the animals. Ambergris costs more than $9,000 per pound and is very difficult to obtain in large quantities; Sensuous Scents can only obtain about 20 pounds of ambergris each year. The other ingredients—deionized water, ethanol, and various additives—are available in unlimited quantities for a reasonable cost.

You have been asked to create a spreadsheet model for Solver to determine the optimal product mix that maximizes Sensuous Scents' net income after taxes.

Setting up the Spreadsheet

The sections in this spreadsheet are different from those in the preceding trucking problem. You will create a Constants section, a Bill of Materials section for the three products, a Quantity Manufactured section that contains the Changing Cells, a Calculations section to calculate ambergris usage, manufacturing costs, and sales revenues per product line, and an Income Statement section to determine the net income after taxes, which will be the Optimization Cell.

AT THE KEYBOARD

Start a new file called **Sensuous Scents Inc.xlsx** and set up the spreadsheet.

Spreadsheet Title and Constants Section

Your spreadsheet title and Constants section should look like Figure D-44. A discussion of the section entries follows the figure.

	A	B	C	D	E	F
1		Sensuous Scents Inc. Product Mix				
2						
3		Constants:		Body Spray	Cologne	Perfume
4		Sales Price per bottle		$11.95	$21.00	$53.00
5		Conversion Cost per Unit (Direct Labor plus Manufacturing Overhead)		$2.60	$6.50	$13.00
6		Minimum Sales Demand		60000	25000	12000
7		Income Tax Rate	0.32			
8		Sales, General and Administrative Expenses per Dollar Revenue	0.30			
9		Available Ambergris (lbs)	20			
10		Cost per lb, Deionized Water	$0.50			
11		Cost per lb, Ethanol	$1.00			
12		Cost per lb, other Additives	$182.00			
13		Cost per lb, Ambergris	$9,072.00			

FIGURE D-44 Spreadsheet title and Constants section for Sensuous Scents Inc.

- Sales Price per bottle—These values are the sales prices for each of the three products.
- Conversion Cost per Unit—These values are the direct labor costs plus the manufacturing overhead costs budgeted per unit manufactured. A conversion cost is often used in industries that manufacture liquid products.
- Minimum Sales Demand—These values reflect the forecast minimum sales demand that you must supply to your customers. These values will be used later as constraints.
- Income Tax Rate—The rate is 32% of your pretax income. No taxes are paid on losses.
- Sales, General and Administrative Expenses per Dollar Revenue—This value is an estimate of the non-manufacturing costs that Sensuous Scents will incur per dollar of sales revenue. These expenses are subtracted from the Gross Profit value in the Income Statement section to obtain Net Income before taxes.
- Available Ambergris (lbs.)—This value is the amount of ambergris that Sensuous Scents obtained this year for production.
- Cost per lb., Deionized Water—This value is the current cost per pound of deionized water.
- Cost per lb., Ethanol—This value is the current cost per pound of ethanol.
- Cost per lb., other Additives—Scent products contain other additives and fixatives to enhance or preserve the fragrance. This value is the cost per pound of the other additives.
- Cost per lb., Ambergris—This value is the current market price per pound of naturally harvested ambergris. Again, no whales were harmed to obtain the ambergris.

The rest of the cells are filled with a gray background to indicate that you will not use their values or formulas. The section is arranged this way to maintain one column per product all the way down the spreadsheet, which will simplify writing the formulas later.

Bill of Materials Section

Your spreadsheet should contain a Bill of Materials section, as shown in Figure D-45. The section entries are explained after the figure. A bill of materials is a list of raw materials and ingredients required to make one unit of a product.

	A	B	C	D	E	F
14						
15		Bill of Materials:		Body Spray	Cologne	Perfume
16		Deionized Water (lb)		0.4	0.1	0.05
17		Ethanol (lb)		0.1	0.02	0.01
18		Other Additives (lb)		0.01	0.001	0.0001
19		Ambergris (lb)		0.0001	0.00018	0.00055

FIGURE D-45 Bill of Materials section

- Deionized Water (lb.)—The amount of deionized water required to make one unit of each product
- Ethanol (lb.)—The amount of ethanol required to make one unit of each product
- Other Additives (lb.)—The amount of other additives required to make one unit of each product
- Ambergris (lb.)—The amount of ambergris required to make one unit of each product

Extremely small quantities of ambergris and other additives are required to make one bottle of each product. Also, each product requires a different amount of ambergris.

Quantity Manufactured (Changing Cells) Section

This model contains a separate Changing Cells section called Quantity Manufactured, as shown in Figure D-46. This section contains the cells that you want Solver to manipulate to achieve the highest net income after taxes.

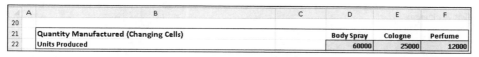

	A	B	C	D	E	F
20						
21		Quantity Manufactured (Changing Cells)			Body Spray	Cologne	Perfume
22		Units Produced			60000	25000	12000

FIGURE D-46   Quantity Manufactured (Changing Cells) section

Cells D22, E22, and F22 are yellow to indicate that Solver will change them to reach an optimal solution. To begin, enter the minimum sales demand in these cells, which will remind you to specify the minimum demand constraints from the Constants section in the Solver Parameters window.

Calculations Section

Your model should contain the Calculations section shown in Figure D-47.

	A	B	C	D	E	F	G
23							
24		Calculations:			Body Spray	Cologne	Perfume	Totals
25		Lbs of Ambergris Used						
26		Manufacturing Cost per Unit (Materials Costs plus Conversion Cost)						
27		Total Manufacturing Costs per Product Line						
28		Sales Revenues per Product Line						

FIGURE D-47   Calculations section

The section contains the following calculations:

- Lbs. of Ambergris Used—This value is the pounds of ambergris per unit from the Bill of Materials section, multiplied by Units Produced from the Quantity Manufactured section for each of the three products. The Totals cell (G25) is the sum of cells D25, E25, and F25. Use the value in this cell to specify the constraint that you only have 20 pounds of ambergris available to use for raw materials (Constants section, cell C9).
- Manufacturing Cost per Unit (Materials Costs plus Conversion Cost)—To get this value, write a formula that multiplies the unit cost for each of the four product ingredients by the amount per unit specified in the bill of materials, multiplied by Units Produced. The total materials costs for the four ingredients are added together, and then the Conversion Cost per Unit is added from the Constants section to obtain the Manufacturing Cost per Unit. Enter the following formula for the Body Spray Manufacturing Cost per Unit in cell D26:

    =C10*D16 + C11*D17 + C12*D18 + C13*D19 + D5

    The Totals cell is not used in this row—you can fill the cell in gray to indicate that it is not used.
- Total Manufacturing Costs per Product Line—This value is the Manufacturing Cost per Unit multiplied by Units Produced from the Quantity Manufactured section. The Totals cell (G27) is the sum of cells D27, E27, and F27. You will use the value in the Totals cell in the Income Statement section.
- Sales Revenues per Product Line—This value is the Sales Price per bottle from the Constants section multiplied by Units Produced from the Quantity Manufactured section. The Totals cell (G28) is the sum of cells D28, E28, and F28. You will use the value in this cell in the Income Statement section.

Income Statement Section

The last section you need to construct is the Income Statement, as shown in Figure D-48. An explanation of the needed formulas follows the figure.

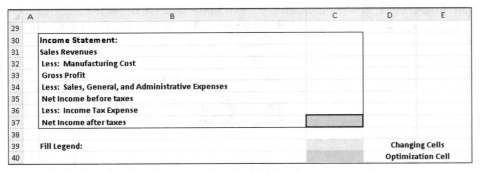

FIGURE D-48   Income Statement section with fill legend

- Sales Revenues—This value is the total sales revenues from the Calculations section (cell G28).
- Less: Manufacturing Cost—This value is the total manufacturing costs from the Calculations section (cell G27).
- Gross Profit—This value is the Sales Revenues minus the Manufacturing Cost.
- Less: Sales, General, and Administrative Expenses—This value is the Sales Revenues multiplied by the Sales, General, and Administrative Expenses per Dollar Revenue from the Constants section (cell C8).
- Net Income before taxes—This value is the Gross Profit minus the Sales, General, and Administrative Expenses.
- Less: Income Tax Expense—If the Net Income before taxes is greater than zero, this value is the Net Income before taxes multiplied by the Income Tax Rate in the Constants section. If Net Income before taxes is zero or less, the Income Tax Expense is zero.
- Net Income after taxes—This value is the Net Income before taxes minus the Income Tax Expense. You will use this value as your Optimization Cell because you want to maximize Net Income after taxes.

Setting up Solver

You need to satisfy the following conditions when running Solver:

- Your objective is to maximize Net Income after taxes (cell C37).
- Your Changing Cells are the Units Produced (cells D22, E22, and F22).
- Observe the following constraints:

  - You must produce at least the Minimum Sales Demand for each product (cells D6, E6, and F6).
  - Your total Lbs. of Ambergris Used (cell G25) cannot exceed the Available Ambergris (cell C9).
  - You cannot produce negative units of any product (enter constraints for the Changing Cells to be greater than or equal to zero).
  - You can only produce whole units of any product (enter constraints for the Changing Cells to be integers).

Run Solver and create an Answer Report when Solver finds the solution. When you complete the program, print your spreadsheet with the Solver solution, and print the Answer Report. Save your work and close Excel.

TROUBLESHOOTING SOLVER

Solver is a fairly complex software program. This section helps you address common problems you may encounter when attempting to run Solver.

Using Whole Numbers in Changing Cells

Before you run your first Solver model or rerun a previous model, always enter a positive whole number in each of the Changing Cells. If you have not already defined maximum and minimum constraints for the values in the Changing Cells, enter 1 in each cell before running Solver.

Getting Negative or Fractional Answers

If you receive negative or fractional answers when running Solver, you may have neglected to specify one or more of the Changing Cells as non-negative integers. Alternatively, if you are working on a cost minimization problem and you fail to specify the Optimization Cell as non-negative, you may receive a negative answer for the cost. Sometimes Solver will also warn you that you have one or more unbounded constraints (see Figure D-49).

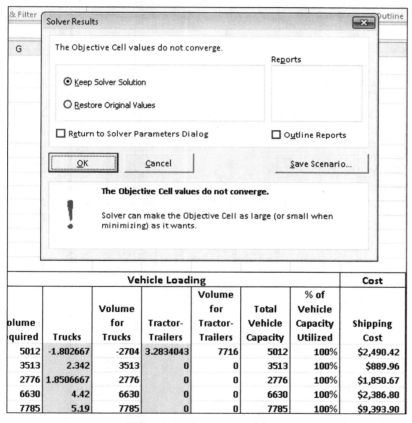

Vehicle Loading							Cost
Volume Required	Trucks	Volume for Trucks	Tractor-Trailers	Volume for Tractor-Trailers	Total Vehicle Capacity	% of Vehicle Capacity Utilized	Shipping Cost
5012	-1.802667	-2704	3.2834043	7716	5012	100%	$2,490.42
3513	2.342	3513	0	0	3513	100%	$889.96
2776	1.8506667	2776	0	0	2776	100%	$1,850.67
6630	4.42	6630	0	0	6630	100%	$2,386.80
7785	5.19	7785	0	0	7785	100%	$9,393.90

FIGURE D-49   Solver has an "unbounded" objective function because you did not specify non-negative integer constraints

Creating Overconstrained Models

If Solver cannot find a solution because it cannot meet the constraints you defined, you will receive an error message. When this happens, Solver may even violate the integer constraints you defined in an attempt to find an answer, as shown in Figure D-50.

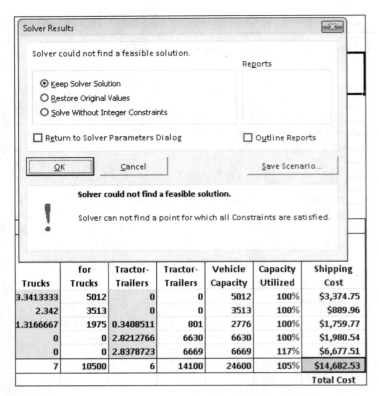

FIGURE D-50   Solver could not find a feasible solution because not enough vehicles were available

Setting a Constraint to a Single Amount

Sometimes you may want to enter an exact amount into a constraint, as opposed to a number in a range. For example, if you wanted to assign exactly 11 trucks in the CV Fitness problem instead of a maximum of 12, you would select the equals (=) operator in the Change Constraint dialog box, as shown in Figure D-51.

FIGURE D-51   Constraining a value to a specific amount

Setting Changing Cells to Integers

Throughout the tutorial, you were directed to set the Changing Cells to integers in the Solver constraints. In many business situations, there is a logical reason for demanding integer solutions, but this approach does have disadvantages. Forcing integers can sometimes increase the amount of time Solver needs to find a feasible solution. In addition, Solver sometimes can find a solution using real numbers in the Changing Cells instead of integers. If Solver cannot find a feasible solution or reports that it has reached its calculation time limit, consider removing the integer constraints from the Changing Cells and rerunning Solver to see if it finds an optimal solution that makes sense.

Restarting Solver with New Constraints

Suppose you want to start over with a completely new set of constraints. In the Solver Parameters window, click Reset All. You will be asked to confirm that you want to reset all the Solver options and cell selections (see Figure D-52).

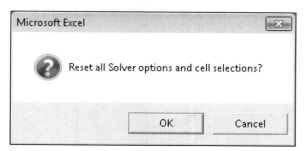

FIGURE D-52   Reset options warning

If you want to clear all the Solver settings, click OK. An empty Solver Parameters window appears with all the former entries deleted, as shown in Figure D-53. You can then set up a new model.

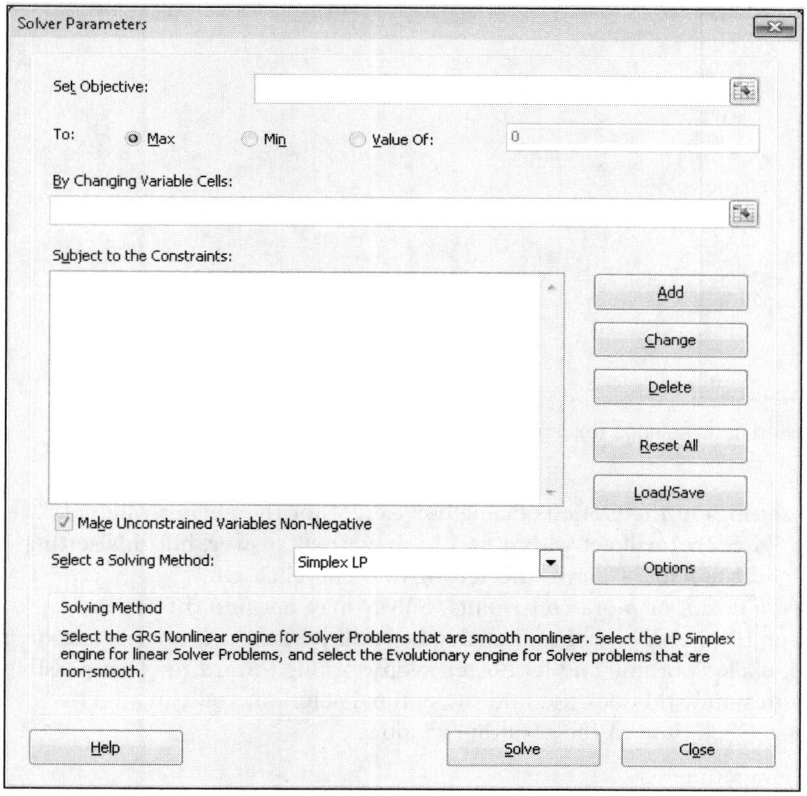

FIGURE D-53   Solver Parameters window after selecting Reset All

Using the Solver Options Dialog Box

Solver has several internal settings that govern its search for an optimal answer. Click the Options button in the Solver Parameters window to see the default selections for these settings, as shown in Figure D-54.

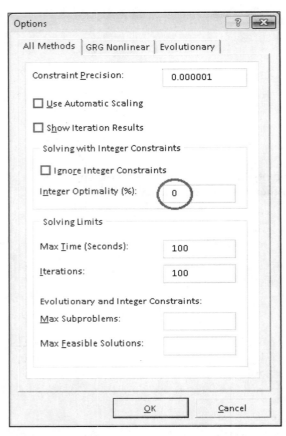

FIGURE D-54   Solver Options dialog box with default settings (except Integer Optimality %)

You should not need to change the settings in the Options dialog box except for the default value of 5% for Integer Optimality. When it is set at 5%, Solver will get within 5% of the optimal answer, but this setting might not give you the lowest cost or highest income. Change the setting to 0 and click OK.

In more complex problems that have a dozen or more constraints, Solver may not find the optimal solution within the default 100 seconds or 100 iterations. If so, a dialog box will prompt you to continue or stop (see Figure D-55). If you have time, click Continue and let Solver keep working toward the best possible solution. If Solver works for several minutes and still does not find the optimal solution, you can stop by pressing the Ctrl and Break keys together. Click Stop in the resulting window.

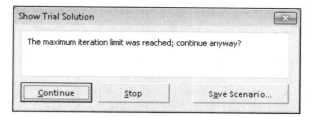

FIGURE D-55   Prompt that appears when Solver reaches its maximum iteration limit

If you think that Solver needs more time and iterations to reach an optimal solution, you can increase the Max Time and Iterations, but you should probably keep both values under 32,000.

Printing Cell Formulas in Excel

Earlier in the tutorial, you learned how to display cell formulas in your spreadsheet cells. Hold down the Ctrl key and then press the ~ key; on most keyboards, this key is near the upper-left corner next to the "1" key. You can change the cell widths to see the entire formula by clicking and dragging the column by the dividing lines between the column letters. See Figure D-56.

	G	H	I	J	K	L
13						
14			Vehicle Loading			
15	Volume Required	Trucks	Volume for Trucks	Tractor-Trailers	Volume for Tractor-Trailers	Total Vehicle Capacity
16	=D16*$C$8+E16*$C$9+F16*$C$10	2	=H16*$C$5	1	=J16*$C$6	=I16+K16
17	=D17*$C$8+E17*$C$9+F17*$C$10	1	=H17*$C$5	1	=J17*$C$6	=I17+K17
18	=D18*$C$8+E18*$C$9+F18*$C$10	2	=H18*$C$5	0	=J18*$C$6	=I18+K18
19	=D19*$C$8+E19*$C$9+F19*$C$10	3	=H19*$C$5	1	=J19*$C$6	=I19+K19
20	=D20*$C$8+E20*$C$9+F20*$C$10	1	=H20*$C$5	3	=J20*$C$6	=I20+K20
21	=SUM(G16:G20)	=SUM(H16:H20)	=SUM(I16:I20)	=SUM(J16:J20)	=SUM(K16:K20)	=SUM(L16:L20)

FIGURE D-56   Spreadsheet with formulas displayed in the cells

To print the formulas, click the File tab and select Print. To restore the screen to its normal appearance and display values instead of formulas, press Ctrl + ~ again; the key combination is actually a toggle switch. If you changed the column widths in the formula view, you might have to resize the columns after you change back.

Solver "Fatal" Errors

When you run Solver, you might sometimes receive a message like the one shown in Figure D-57.

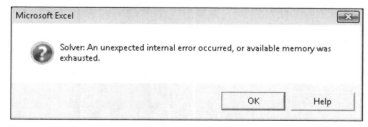

FIGURE D-57   Solver "fatal" error

Solver usually attempts to find a solution or reports why it cannot. When Solver reports a fatal error, the root cause is difficult to troubleshoot. Possible causes include merged cells on the spreadsheet or printing multiple Answer Reports after running Solver multiple times. In Excel 2007, a common solution to this error was to remove the Solver add-in, close Excel, reopen it, and then reinstall Solver. If you encounter a fatal error when using this book, check with your instructor.

Sometimes Solver will generate strange results. Even when your cell formulas and constraints match the ones your instructor has created, Solver's answers might not match the "book" answers. You might have entered your constraints into Solver in a different order, you may have changed some of the options in Solver, or you may have specified real numbers instead of integers for the constraints (or vice versa). Also, the solving method you selected and the amount of time you gave Solver to work can affect the final answer. If your solution is close to the one posted by your instructor, but not exactly the same, show the instructor your setup in the Solver Parameters window. Solver is a powerful tool, but it is not infallible—ask your instructor for guidance if necessary.

KUHLMAN'S DEPARTMENT STORE MEDIA PROBLEM

Decision Support Using Excel Solver

PREVIEW

Kuhlman's Department Store is a family-owned business in a thriving Midwestern city. The store has been successful for more than 90 years, but recently it has come under serious pressure from a new suburban shopping mall. Sales and operating income have been falling in the past two years since the mall opened, and the store is now barely making a profit.

The owners recently hired a media consulting firm to evaluate the store's advertising strategy and suggest improvements. Kuhlman's has traditionally used direct mail and newspaper advertisements, but the media consultants recommended that the store consider TV, radio, and magazine ads, and even Internet ads for advertising alternatives. The consultants performed a media study and developed estimates for potential audience size, cost per ad, recommended maximum number of ads, and the estimated percentage of customers gained from the added media exposure.

You are the MIS manager for Kuhlman's, and you have been asked to build a spreadsheet model to determine the best mix of media to optimize the store's net income, given an advertising budget of $80,000.

PREPARATION

- Review the spreadsheet concepts discussed in class and in your textbook.
- Your instructor may assign Excel exercises to help prepare you for this case.
- Tutorial D explains how to set up and use Solver for maximization and minimization problems.
- Review the file-saving instructions—it is always a good idea to save an extra copy of your work on a USB thumb drive.
- Reviewing Tutorial F will help you brush up on your presentation skills.

BACKGROUND

You have been asked to use your Excel skills to build a decision model and determine how much of each advertising medium Kuhlman's should purchase in order to maximize the company's net income. To build this model, you need the following data, which the media consultants compiled:

- The six advertising media to consider—Ads for the Internet, television, radio, newspaper, magazines, and direct mail
- The potential audience size in the region that Kuhlman's serves
- The cost per ad, cost per commercial, or cost per Web site hit (for Internet ads)
- The maximum amounts of advertising for each available medium
- The percentage of customer visits expected as a result of each ad or commercial

In addition, the Accounting Department has given you the following information about the operating budget based on historical revenues and costs:

- The average sales per new customer per year
- The maximum budget for advertising expenses

- The estimated cost of goods sold as a percentage of sales
- Other operating expenses, exclusive of the advertising expenses
- The current corporate income tax rate

Your Solver model will calculate the total advertising expenses for any or all of the six media evaluated, the number of customers you expect from the advertising media selected, the expected sales from those customers, and the cost of goods sold from the expected sales. You will use the results of these calculations to create an income statement that ends with net income after taxes; the Solver model will optimize this cell.

Advertising Media Considered

For the purposes of this case, the types of advertising media and their cost structures are greatly simplified. In actual practice, the pricing schedules for each type of advertising vary depending on the amount of advertising purchased, the dates and times that TV or radio commercials air and that newspaper ads run, the size and placement of newspaper and magazine ads, and the pricing structure and search engines offered for Internet ads. You can leave those decisions to the marketing consultants.

This section briefly describes each type of marketing medium under consideration:

- Internet ads—An increasingly popular method of attracting new customers, Internet ads target e-commerce customers, but they can also attract customers to "brick and mortar" stores as well. The power of Internet ads is that they can be linked to popular Web search engines such as Google, Yahoo, and Bing!, generating millions of contacts or "hits." The pricing structure for Internet ads is highly variable—the pricing selected for this case is "pay per click," in which a company pays pennies or even fractions of a penny to a Web advertiser for each Web user who clicks the advertisement link.
- Television—Since their integration into American culture more than 60 years ago, television commercials have become the predominant marketing medium. The combination of moving picture and audio can convey a powerful message to potential buyers. Although television commercials can be expensive to produce and air, and the message is brief (a 30-second Super Bowl ad costs $2.5 million!), the cost per potential customer contacted is reasonable.
- Radio—What radio commercials lack in visual impact, they make up for in portability. A large proportion of the working population listens to radio in cars during the daily commute to and from work. An additional appeal of radio is that listeners tend to be loyal to one or two local stations rather than "surfing channels" as television viewers do. Radio ads cost less to produce and air than television advertising.
- Newspaper ads—The oldest of the marketing media, newspaper ads offer visual impact while allowing readers unlimited time to peruse the ad. The emergence of modern media and changing demographics in America have degraded the media value of many newspapers to the point that they are now largely collections of advertisements with surprisingly little news. Despite this decline, a large proportion of older Americans (who are also loyal department store customers) continue to read newspapers. By purchasing ads in the local newspapers, Kuhlman's has been targeting its traditional customer base.
- Magazine ads—The advantages and disadvantages of newspaper advertising also apply to magazine ads. One exception is that topical magazines offer targeted audiences to businesses offering specific products. For example, fashion magazines feature clothing ads, and "home and garden" magazines are filled with home improvement ads. Although it would make no sense for a local department store to advertise in a national magazine, Kuhlman's could choose to buy ads in a local magazine.
- Direct mail—Despite growing dissatisfaction among consumers with "junk mail," the chief advantage of sending direct mail such as catalogs is that a company can target its existing customer base for repeat sales and display a larger selection of merchandise in its mailer. Of all the advertising media, direct mail has the highest cost per customer contacted, largely due to postage. The bulk of Kuhlman's past advertising budgets has been spent on direct mail.

ASSIGNMENT 1: CREATING A SPREADSHEET FOR DECISION SUPPORT

In this assignment, you will create spreadsheets that model the business decision Kuhlman's is seeking. In Assignment 1A, you will create a Solver spreadsheet to model the media selection decision. In Assignment 1B, you will copy and rerun the Solver spreadsheets, given changes in media costs for television and magazine ads. In Assignments 2 and 3, you will use the spreadsheet models to develop recommendations for the best media mix, and you will prepare an oral presentation of your analysis and recommendations.

This section helps you set up each of the following spreadsheet components before entering the cell formulas:

- Constants
- Calculations
- Income Statement

The Calculations section is the heart of the decision model. You will set up each of the media in its own column; one of the rows will be the Changing Cells range for Solver to manipulate. The final Net Income after taxes in the Income Statement will serve as your optimization cell.

> **NOTE**
>
> You cannot use an IF statement for the income tax calculation if you want to use the Simplex LP method to solve the decision model. When used to calculate the income tax, the IF statement will render the decision model nonlinear and require a different optimization method in Solver.

Assignment 1A: Creating the Spreadsheet—Base Case

A discussion of each spreadsheet section follows. This information will help you set up each section of the model and learn the logic of the formulas in the Calculations and Income Statement sections. If you choose to enter the data directly, follow the cell structure shown in the figures. *You can also download the spreadsheet skeleton if you prefer.* To access the base spreadsheet skeleton, go to your data files, select Case 8, and then select **Kuhlmans.xlsx**.

Constants Section

First, build the skeleton of your spreadsheet. Set up the spreadsheet title and Constants section as shown in Figure 8-1. An explanation of the line items follows the figure.

	A	B	C	D	E	F	G	H	I
1			Kuhlman's Department Store--Media Problem						
2									
3		Constants							
4		Average Sales per new customer per year	$100						
5		Maximum Budget for Advertising	$80,000						
6		Cost of Goods as a percent of Sales	80%						
7		Other Operating Expenses	$820,000						
8		Income Tax Rate	22%						

FIGURE 8-1   Spreadsheet title and Constants section

- Spreadsheet title—Enter the spreadsheet title in cell B1, and then merge and center the title across cells B1 through I1. In Figure 8-1, the title font is 14-point Arial bold. The rest of the spreadsheet uses Arial bold as well; section titles are 12 points, and the rest of the spreadsheet is 11-point text.
- Average Sales per new customer per year—This value, $100.00, is the average sales per new customer per year that Kuhlman's has determined from its historical records.
- Maximum Budget for Advertising—This value, $80,000, is the amount of money that Kuhlman's has budgeted this year for advertising expenses, exclusive of the fee that was paid to the media consultants. Usually this expense is part of the Other Operating Expenses, but it has been broken out separately to use in this model as a constraint.

- Cost of Goods as a percent of Sales—This value, 80%, is the amount that Kuhlman's paid to purchase the merchandise it sells, expressed as a ratio of its sales revenue. The cost of goods also includes the costs of freight-in, duties, and preparing the merchandise for sale.
- Other Operating Expenses—This value, $820,000, is Kuhlman's budgeted operating expenses for the year, exclusive of the advertising budget.
- Income Tax Rate—This value, 22%, is the current income tax rate.

Calculations Section

Your Calculations section will contain the data obtained from the media consultants for each of the advertising media. It will also contain the Changing Cells row, the calculations for Advertising Expense, Customers, Sales, and Cost of Goods, and the totals for all the calculations. See Figure 8-2. An explanation of the line items follows the figure.

	A	B	C	D	E	F	G	H	I
9									
10		Calculations							
11		Media Considered	Internet Ads	TV	Radio	Newspaper	Magazine	Direct Mail Ads	
12		Potential Audience	3,000,000	400,000	250,000	45,000	55,000	250,000	
13		Cost per Ad or Web Hit	$0.010	$3,000	$1,000	$200	$500	$0.40	
14		Number of Ads or Web Hits	0	0	0	30	0	185000	
15		Maximum Ads or Web Hits Available	3000000	16	30	52	26	250000	
16		% of customers	0.50%	0.70%	0.80%	1.50%	0.80%	14.00%	Totals
17		Advertising Expense							
18		Customers							
19		Sales							
20		Cost of Goods							
21		Sales per Advertising Dollar							

FIGURE 8-2 Calculations section

- Media Considered—This line contains the titles for each of the six media types, which you enter into cells C11 through H11: Internet Ads, TV, Radio, Newspaper, Magazine, and Direct Mail Ads.
- Potential Audience—These values are the average audience in the region exposed to one airing or printing of the advertisement for each media type. Enter the following values into cells C12 through H12 in the order shown: 3,000,000, 400,000, 250,000, 45,000, 55,000, and 250,000.
- Cost per Ad or Web Hit—These values are the cost of an ad in each advertising medium. Note that Internet ads are charged "per click" or per Web hit to the ad site. Enter the following values into cells C13 through H13 in the order shown: $.010, $3,000, $1,000, $200, $500, $.40.
- Number of Ads or Web Hits—This line is the heart of the Solver model: the Changing Cells. These values are the amounts of each type of advertising that Solver recommends for Kuhlman's. Enter a zero in cells C14, D14, E14, and G14, enter 30 in cell F14 for Newspaper Ads, and enter 185,000 in cell H14 for Direct Mail Ads. These values represent Kuhlman's historical spending for advertising (see the following Note regarding Internet ads). You might want to fill in the cells with a background color to indicate that they will be the Changing Cells for Solver.
- Maximum Ads or Web Hits Available—These cells are the limits for the amounts of ads purchased, as recommended by the media consultants. You will use these cells to define constraints for the amount of each medium purchased. Enter the following values into cells C15 through H15 in the order shown: 3,000,000, 16, 30, 52, 26, 250,000.
- % of customers—These cells contain estimates of the percentage of new customers obtained from exposure to each type of advertising media. The media consultants were conservative in their estimates for the impact of all media types except direct mail ads. Kuhlman's provided the figure of 14.00% based on customers who brought in direct mail ads for a special discount during the past year. Enter the following values into cells C16 through H16 in the order shown: .50%, .70%, .80%, 1.50%, .80%, and 14.00%.

NOTE—INTERNET ADS AND WEB "HITS"

Kuhlman's cannot actually dictate how many Web ads it will pay for, other than specifying a maximum charge that depends on the number of potential customers who click the Web ad. This advertising medium was included in the case to recognize the potential of the Internet to attract customers. In the case of a local department store, the impact of Internet advertising may not be significant.

The calculations that make up the rest of the section are described next:

- Advertising Expense—This value is the Cost per Ad or Web Hit multiplied by the Number of Ads or Web Hits.
- Customers—This value is the estimated number of customers gained from each of the media. The calculation for this value is different for the Internet Ads and Direct Mail Ads than it is for the rest of the media. The customers gained from the Internet Ads and Direct Mail Ads is simply the Number of Ads or Web Hits multiplied by the % of customers. For the other four media (TV, radio, newspapers, and magazine), the customers gained equals the Potential Audience multiplied by the Number of Ads multiplied by the % of customers. Double-check your formulas to make sure you entered them correctly.
- Sales—This value is the number of customers gained multiplied by Average Sales per new customer per year. The latter value is taken from the Constants section.
- Cost of Goods—This value is the Sales multiplied by the Cost of Goods as a percent of Sales. The latter value is taken from the Constants section.
- Sales per Advertising Dollar—This value is the Sales divided by the Advertising Expense. Although this value is not needed to solve the problem, it will show the management team the relative effectiveness of each advertising medium purchased. You should use an IF statement to place "NA" or another appropriate comment in the cell if the Advertising Expense is zero. Solver might not select certain advertising media. If you do not use the IF statement and there is no Advertising Expense, the cell will show a division by zero error.
- Totals (for Advertising Expense, Customers, Sales, and Cost of Goods)—You should total each of these rows and enter the results in cells I17, I18, I19, and I20, respectively. The total number of new customers is not strictly necessary, but the other three totals are needed for further calculations or constraints.

If you wrote the formulas correctly, your Calculations section should look like Figure 8-3.

	B	C	D	E	F	G	H	I
9								
10	**Calculations**							
11	**Media Considered**	Internet Ads	TV	Radio	Newspaper	Magazine	Direct Mail Ads	
12	Potential Audience	3,000,000	400,000	250,000	45,000	55,000	250,000	
13	Cost per Ad or Web Hit	$0.010	$3,000	$1,000	$200	$500	$0.40	
14	Number of Ads or Web Hits	0	0	0	30	0	185000	
15	Maximum Ads or Web Hits Available	3000000	16	30	52	26	250000	
16	% of customers	0.50%	0.70%	0.80%	1.50%	0.80%	14.00%	Totals
17	Advertising Expense	$0.00	$0	$0	$6,000	$0	$74,000	$80,000
18	Customers	0	-	-	20,250.0	-	25,900.0	46,150.0
19	Sales	$0	$0	$0	$2,025,000	$0	$2,590,000	$4,615,000
20	Cost of Goods	$0	$0	$0	$1,620,000	$0	$2,072,000	$3,692,000
21	Sales per Advertising Dollar	NA	NA	NA	$338	NA	$35	

FIGURE 8-3 Completed Calculations section—historical data

Income Statement Section

Kuhlman's income statement (see Figure 8-4) is actually a projection for the coming year, and is based either on its traditional advertising mix or the solution you produce with Solver. This section explains how you should structure the Income Statement section and explains the formulas for each cell.

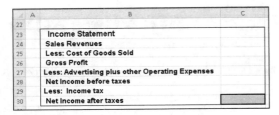

FIGURE 8-4 Income Statement section

- Sales Revenues—This value is the total Sales revenues from the Calculations section (cell I19).
- Less: Cost of Goods Sold—This value is the total Cost of Goods from the Calculations section (cell I20).
- Gross Profit—This value is the Sales Revenues minus the Cost of Goods Sold.
- Less: Advertising plus other Operating Expenses—This value is the sum of the total Advertising Expense from the Calculations section (cell I17) and the Other Operating Expenses from the Constants section (cell C7).
- Net Income before taxes—This value is the Gross Profit minus the Advertising plus other Operating Expenses.
- Less: Income tax—If the Net Income before taxes is greater than zero, then the store made a profit, and this value is the Net Income before taxes multiplied by the Income Tax Rate from the Constants section. If the store makes nothing or has a net loss, then the income tax is zero. However, assume that the store made a profit. Using an IF statement in this line of calculations will create a nonlinear model, which means you will not be able to use the Simplex LP solving method. Later in the case, however, you can modify this calculation with the IF statement and use a different solving method.
- Net Income after taxes—This value is the Net Income before taxes minus the income tax. You will optimize this cell because you want to maximize Net Income after taxes in this case, so you should fill the cell with an appropriate background color to easily identify it as the Optimization Cell.

If your formulas are correct, the "historical" Income Statement for Kuhlman's will appear as shown in Figure 8-5.

	A	B	C
22			
23		Income Statement	
24		Sales Revenues	$4,615,000
25		Less: Cost of Goods Sold	$3,692,000
26		Gross Profit	$923,000
27		Less: Advertising plus other Operating Expenses	$900,000
28		Net Income before taxes	$23,000
29		Less: Income tax	$5,060
30		Net Income after taxes	$17,940

FIGURE 8-5 Completed Income Statement for historical advertising

As you can see, Kuhlman's is barely making money, which is why it is considering other advertising media.

Setting up and Running Solver

Before using the Solver Parameters window, you should consider jotting down the parameters you must define and their cell addresses. Here is a suggested list:

- The cell you want to optimize (Net Income after taxes) and whether you want to minimize or maximize it
- The cells you want Solver to manipulate to obtain the optimal solution (Changing Cells)
- The constraints you will define:
  - Each Changing Cell has a maximum amount of advertising you can buy (in the row below the Changing Cells).

- Each Changing Cell must be an integer greater than or equal to zero—you cannot buy half an ad.
- Your total spending on advertising cannot exceed the limit specified in the Constants section.

Next, you should make a copy of your "historical" worksheet to use with Solver. When Solver finds a solution, it will replace the historical advertising media selections with the new optimized selections. To create a copy of the spreadsheet in the same workbook, right-click the tab at the bottom of the worksheet. In Figure 8-6, note that the worksheet is named Kuhlmans—Historical. Select Move or Copy from the menu.

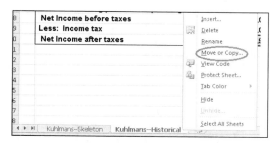

FIGURE 8-6 Selecting the Move or Copy command

When the Move or Copy dialog box appears (see Figure 8-7), click Create a copy, click the (move to end) option, and then click OK.

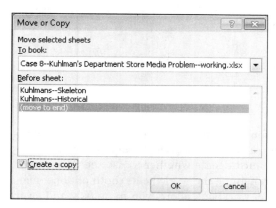

FIGURE 8-7 Move or Copy worksheet dialog box

The new copy of the worksheet will appear with the old worksheet's name followed by a *(2)*. Right-click the name tab to rename the worksheet (see Figure 8-8). In this example, the worksheet was renamed Kuhlmans—Solver.

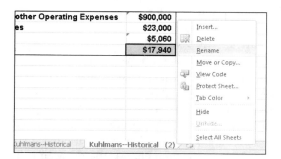

FIGURE 8-8 Renaming the copied worksheet

Next, go to the Analysis group in the Data tab and click Solver to set up your problem (see Figure 8-9). Use the Simplex LP solving method.

FIGURE 8-9   The Solver Parameters window

Run Solver and ask for the Answer Report when Solver has found a solution that satisfies the constraints. When you finish, print the entire workbook, including the Solver Answer Report Sheet. Save the workbook by clicking the File tab and then Save. For the rest of the case, you must decide whether to use the Save As command for new Excel workbooks or to continue copying and renaming the worksheets. Both options offer distinct advantages, but having all of your worksheets in one Excel workbook allows you to quickly and easily compare different solutions as well as prepare summary reports.

Before continuing, look at the advertising mix that Solver chose for optimizing the net income. If you set up Solver correctly, you should see a dramatic difference in the current net income and the net income from the media selection that Solver made. Remember that these numbers are only estimates—actual sales increases will depend on other factors as well, including the state of the economy, the quality and timing of the ads, and what the competition does. Even with these factors included, however, the media consultants and your decision support model have provided Kuhlman's with useful financial data.

Assignment 1B: Changing Parameters and Creating New Solutions

Using the model you created in Assignment 1A, Kuhlman's can also review figures for different media providers and see the effects of negotiating reduced ad prices on the decision model. Now that you know how to copy spreadsheets, you can easily create additional spreadsheets and look at the mix that Solver recommends, given changes in the media pricing. You will examine two reductions in ad prices: First, what happens if the television station reduces its price per commercial from $3,000 to $2,500? Next, with the television discount included, you will rerun Solver with magazine ads reduced from $500 to $300. Copy your original Solver solution worksheet to a new worksheet named **Kuhlmans—TV Discount**, then change the value in cell D13 (TV media cost) from $3,000 to $2,500. When Solver reaches its solution, create an Answer Report—it will be named Answer Report 2 automatically. Examine the report to see how the TV discount changed your media mix and net income. Next, copy the TV Discount worksheet and rename it **Kuhlmans—Mag Discount**. Change the value in cell G13 from $500 to $300, and rerun Solver. When Solver reaches its solution, create another Answer Report, which will be named Answer Report 3 in a new tab. Again, examine the results to see how the discounted magazine ad changed your media mix and net income.

The last change you will examine is the effect of changing the model's solving method from Simplex LP to Evolutionary. Copy the worksheet that contains the original Solver solution data, and rename the new worksheet **Kuhlmans—Evolutionary**.

The Evolutionary solving method can work with nonlinear, "non-smooth" problems, such as those that contain IF statements in their formulas. If you recall, you could not use an IF statement to calculate the income tax because the Simplex LP method will not work with it. In cell C29 of the Kuhlmans—Evolutionary worksheet, enter the IF statement needed to calculate the income tax. Remember that if the Net Income before taxes is less than zero, the income tax is zero. Otherwise, the income tax is the Net Income before taxes multiplied by the Income Tax Rate from the Constants section. To demonstrate that the Simplex LP method will not work, select the method in the Solver Parameters window, and then attempt to run Solver. You will see an error message like the one in Figure 8-10.

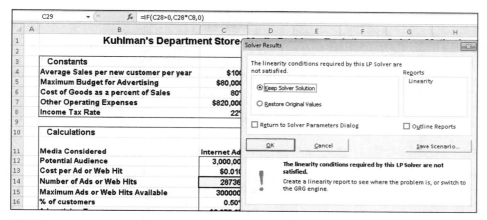

FIGURE 8-10   Error message when Simplex LP method is used on nonlinear models

In Excel 2010, Solver can use three solving methods: Simplex LP, GRG Nonlinear, and Evolutionary. You can consult Microsoft Online Help for an explanation of each method. For most Solver decision models, Simplex LP works well, but as you have seen, it probably will not work for nonlinear models. You could try the GRG Nonlinear method, but in this case it would not work either. Therefore, open the Solver Parameters window and change the solving method to Evolutionary. The method will take longer to solve the problem than the Simplex LP method, but you should get a solution. Create another Answer Report, then compare the Income Statement from the Evolutionary solution with that of your original Solver solution using Simplex LP. The net income in the Evolutionary solution will probably be $10,000 to $20,000 less, depending on the amount of time you gave the Evolutionary model to run the problem. To summarize, if you can keep your formulas linear (no IF formulas or second-order equations), Simplex LP is the best solving method.

Modify the titles at the top of each worksheet to append an appropriate description. For instance, the original historical worksheet could be titled **Kuhlman's Department Store—Media Problem—Historical**. When you finish modifying the titles, save your workbook and print all of your new worksheets and Answer Reports.

ASSIGNMENT 2: USING THE WORKBOOK FOR DECISION SUPPORT

You have built a series of worksheets to determine the advertising media mix that optimizes Kuhlman's net income while staying within the prescribed advertising budget. You will now complete the case by using your solutions and Answer Reports to gather the data needed to make the advertising decisions and by documenting your recommendations in a memorandum.

Assignment 2A: Using Your Workbook to Gather Data

Use your consolidated workbook and the printouts you created in Assignment 1B to create a summary table of results that you will place in your memo. The easiest way to make this table is to create a new worksheet in your workbook, and then copy and paste the new table into your memo. The table format is shown in Figure 8-11.

Model Examined:	Historical	Solver	TV Ad Discount	TV & Magazine Ad Discounts
Internet Ads				
TV Commercials				
Radio Commercials				
Newspaper Ads				
Magazine Ads				
Direct Mail Ads				
Total Advertising Expense				
Total Expected Sales				
Expected Net Income after Taxes				

Kuhlman's Department Store Advertising Analysis--Summary of Results

FIGURE 8-11 Format of table to be used in memo

A 3-D column chart of the expected net income will add visual impact to your presentation. You can insert the chart in the same worksheet as the table, as shown in Figure 8-12. To start a chart, select your four model headings in the table (cells C4 to F4), and then hold down the Ctrl key to select the four Expected Net Income after Taxes cells (cells C13 to F13). Next, click the Insert tab on the Ribbon. In the Charts group, click Column, and then select 3-D Column from the menu.

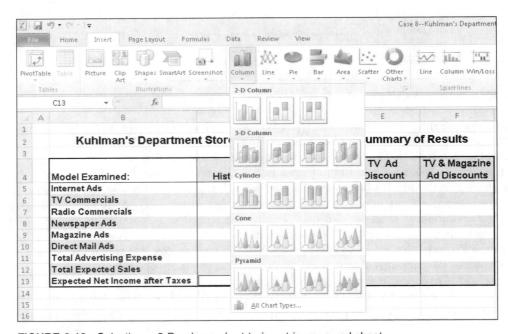

FIGURE 8-12 Selecting a 3-D column chart to insert in your worksheet

A 3-D column chart is placed in the worksheet. Click it to select it, and then reposition the chart beneath your table. Enter an appropriate title for the chart. You can watch the chart values become columns as you fill in the Expected Net Income after Taxes cells (see Figure 8-13).

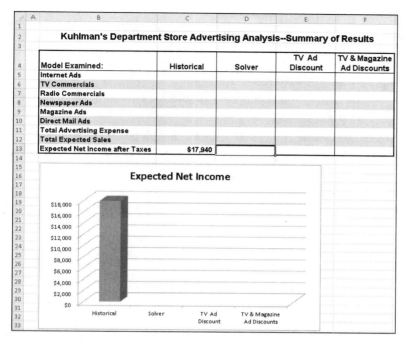

FIGURE 8-13   Building a 3-D column chart while populating your table

Assignment 2B: Documenting Your Recommendations in a Memo

Use Microsoft Word to write a brief memo to the CEO of Kuhlman's Department Store, Harold Kuhlman. State the results of your analysis and recommend how much of each advertising medium to purchase. From your analysis, you should have learned that Kuhlman's must change its marketing strategy from direct mail to be successful. Your memo should observe the following requirements:

- Set up your memo as described in Tutorial E.
- In the first paragraph, briefly describe the situation and state the purpose of your analysis.
- Next, summarize the results of your analysis and state your recommended action.
- Support your recommendation with your Summary of Results table copied from the Excel workbook. (For a description of how to copy and paste Excel objects, see Tutorial C.)
- You should also copy your chart from the Excel workbook into the memo for maximum visual impact.
- In a future case, you might suggest revisiting the Solver analysis at the end of the year to compare Kuhlman's actual sales performance with the estimates from the model. If the advertising mix decisions are successful, Kuhlman's might want to analyze the effect of increasing its advertising budget.

ASSIGNMENT 3: GIVING AN ORAL PRESENTATION

Your instructor may request that you give your analysis and recommendations in an oral presentation. If so, assume that the CEO wants the presentation to last 10 minutes or less. When preparing your presentation, use visual aids such as PowerPoint slides or handouts that you think are appropriate. Tutorial F explains how to prepare and give an effective oral presentation.

DELIVERABLES

Prepare the following deliverables for your instructor:

1. A printout of the memo
2. Printouts of all your worksheets and Answer Reports
3. Your Word document, Excel workbook, and PowerPoint presentation on electronic media or sent to your course site as directed by your instructor

Assemble your paper documentation neatly and staple it together with the memo on top. If you used more than one Excel workbook (.xlsx) file for your case, include a note for your instructor that describes the different files. Good luck!

CASE **9**

GREEN JEANS, LLC SALES AND OPERATIONS PLAN

Decision Support Using Excel Solver

PREVIEW

Green Jeans, LLC is a blue jeans manufacturer in the Southwestern United States. Kathy Green, president of Green Jeans, founded the company in 2005 with a unique vision—to be a competitive, eco-friendly maker of high-quality blue jeans. Green Jeans buys fabric made only from organically grown cotton. It uses wind farm and solar panel electricity to run its plant processes (mostly sewing machines), it uses evaporative cooling in the plant to reduce air conditioning requirements, and through extensive use of internal recycling for raw materials and packaging, Green Jeans has become the first "zero-waste" manufacturer in the clothing industry.

Green Jeans is located in a region that has many textile and clothing mills, a highly mobile workforce, and a highly variable unemployment rate. Textile workers are always looking for jobs, and many clothing mills in the area routinely hire and lay off both full-time and part-time labor. Green Jeans has followed similar hiring and layoff patterns depending on the sales demand for its jeans.

You are the MIS manager for Green Jeans, and have been asked to participate in the semiannual Sales and Operations planning session. Specifically, you must develop and evaluate different sales and operations plans based on widely varying business objectives.

PREPARATION

- Review the spreadsheet concepts discussed in class and in your textbook.
- Your instructor may assign Excel exercises to help prepare you for this case.
- Tutorial D covers how to set up and use Solver for maximization and minimization problems.
- Review the file-saving instructions—it is always a good idea to save an extra copy of your work on a USB thumb drive.
- Reviewing Tutorial F will help you brush up on your presentation skills.

BACKGROUND

You will use your Excel skills to build a decision model for determining the best combination of employment and inventory levels to minimize costs for Green Jeans' Sales and Operations plan (S&OP), which covers the first six months of 2012. Most of these plans typically cover a full business year, but Kathy Green has decided to run a shorter six-month "moving plan," in which the company accountant enters the "Actuals" data at the end of every month, then moves that month into historical data while adding another month to the plan.

S&OP plans vary greatly in manufacturing businesses, but all of them contain at least three sections:

1. The sales forecast, which is expressed both in units sold and sales dollars, and which includes an Actuals cell to enter the actual sales at the end of the month.
2. The operations plan, expressed in units to produce and the resources required to produce them. In the case of Green Jeans, the primary resource is manpower.
3. The inventory plan, expressed as Month End Finished Goods Inventory in units. When the company produces more units than sales in a given period, the finished goods inventory increases. When sales exceed units made in a given period, the finished goods inventory decreases.

At your first meeting to create the S&OP, you quickly discover that key members of management have been at odds over the best production, hiring, and inventory strategies to employ. The argument revolves around three key managers:

- George Detweiler, the Operations manager, feels that the best strategy is to make exactly what the sales forecasts dictate every month, and to hire and lay off employees accordingly. In sales and operations planning, this is called "chasing" the sales forecast or a Chase plan.
- Betty Miller, the Human Resources (HR) manager, disagrees with George. She thinks that the best strategy is to find the correct workforce size to keep employment levels stable, and to let finished goods inventories expand and contract as a buffer against high or low demand. She believes the savings that result from not continually hiring and laying off employees will more than offset the higher costs of carrying inventory. She also contends that a steady, trained workforce will improve product quality and morale in the workplace. (She will also have less work to do if the company hires and trains fewer new employees.) This type of S&OP is called a Level plan.
- Pedro Sanchez, the Logistics manager, disagrees with George and Betty. He contends that the best way to plan operations is to keep the lowest possible amount of finished goods inventory in the warehouse needed to service the customers. He feels that the S&OP should target a 5- to 6-day inventory at the end of every month to minimize inventory carrying costs. This type of S&OP is called a Logistics Target plan.
- Kathy Green, the president, is not sure which strategy to pursue. She is leaning toward Betty Miller's proposal because she believes that minimizing layoffs will build a loyal, well-trained workforce. At the same time, Kathy recognizes the mobility of the textile labor market in the region and does not want to lose a competitive edge from incurring higher inventory carrying costs.

You want to provide quantitative data that shows the management team the most cost-effective solution. You are not sure that any of the managers has the correct solution, so you have decided to build a decision support model to determine the costs, workforce, and inventory levels for each of the managers' proposals. You also want to determine the least expensive optimal solution.

The Accounting Department has given you information for the historical performance of Green Jeans, as well as data that will help you build your model. The information includes:

- Historical sales and operations results for the previous three months
- Employee productivity measured in pairs of jeans produced per employee per day
- Hiring and layoff costs per hire or layoff
- Inventory carrying costs as a percentage of inventory value
- The budgeted unit cost and current selling price of a pair of jeans

The Marketing and Sales Department has provided its sales forecast for the first six months of 2012. Human Resources has provided the current number of employees. The Logistics manager said that to keep customer service at an acceptable level, the end-of-month finished goods inventory must equal at least five days of supply according to the sales forecast.

ASSIGNMENT 1: CREATING A SPREADSHEET FOR DECISION SUPPORT

In this assignment, you will create a spreadsheet that models Green Jeans' S&OP, with an emphasis on employment levels for the six months in question. In Assignment 1A, you will create a Solver spreadsheet to model George Detweiler's Chase plan. In Assignment 1B, you will copy the first spreadsheet to new sheets and rerun Solver to model the HR manager's plan, the Logistics manager's plan, and your own optimal cost plan. In Assignments 2 and 3, you will use the spreadsheet models to summarize the costs of each plan, and you will give a presentation of your analysis and recommendations.

The Sales and Operations spreadsheet model will contain the following sections:

- Constants
- Sales and Operations Plan, which is a combination of the Changing Cells and some calculations
- Hires and Layoffs, which is a calculation of the number of workers hired and laid off
- Costs, which is a calculation and summation of the hiring, layoff, and inventory costs; the Total Cost is the Optimization Cell

Instead of a unified Calculations section, this model setup has calculations spread throughout the last three sections. A typical Sales and Operations plan does not include the hiring levels and cost sections shown in this model, but instead displays them as separate worksheets or reports linked to the plan. Sales and Operations planning modules in large, integrated software packages usually create separate reports of hiring and cost calculations. In this case, you will keep everything on one worksheet.

Assignment 1A: Creating the Spreadsheet—Base Case

A discussion of each spreadsheet section follows. It will help you set up each section of the model and learn the logic of the formulas. If you choose to enter the data directly, follow the cell structure shown in the figures. *To save time, it is highly recommended that you use the spreadsheet skeleton.* To access the base spreadsheet skeleton, go to your data files, select Case 9, and then select **Green Jeans SOP Skeleton.xlsx.**

Constants Section

First, build the skeleton of your spreadsheet. Set up your spreadsheet title and Constants section, as shown in Figure 9-1. An explanation of the line items follows the figure.

	A	B	C	D	E	F	G	H	I	J	K	L
1				Green Jeans, LLC Sales and Operations Plan Jan-Jun 2012								
2												
3		Constants										
4		Employee Productivity	20	units/worker-day								
5		Hiring Cost	$200	per worker								
6		Layoff Cost	$500	per worker								
7		Inventory Carrying Cost	2%	per month applied to ending monthly inventory								
8		Minimum Inventory Level	5	days of supply								
9		Beginning labor force	550	workers								
10		Unit Cost of Inventory	$20									
11		Sales Price per unit	$30									

FIGURE 9-1   Spreadsheet title and Constants section

- Spreadsheet title—Enter the spreadsheet title in cell B1, and then merge and center the title across cells B1 through K1. In Figure 9-1, the title font is 14-point Arial bold. The rest of the spreadsheet uses Arial bold as well; section titles are 12 points, and the rest of the spreadsheet uses 11-point text.
- Employee Productivity—This value, 20, is the number of pairs of jeans (units) produced per worker each day.
- Hiring Cost—This value, $200, is the amount of money required for Green Jeans to hire and train one worker.
- Layoff Cost—This value, $500, is the amount of money required to lay off one worker. This cost includes severance pay and unemployment insurance copay to the state.
- Inventory Carrying Cost—This value, 2%, is the monthly cost of holding the finished goods inventory as a percentage of its unit cost to make.
- Minimum Inventory Level—This value is the minimum inventory level needed to satisfy sales demand, expressed as a number of days. This value determines the service level for customer orders; that is, Green Jeans must keep enough jeans in inventory to have the sizes and quantities needed to fill orders. To maintain a service level of 98%, Green Jeans needs to stay at or above five days of inventory for sales demand.
- Beginning labor force—This value, 550, is the number of workers employed at Green Jeans at the start of the year.
- Unit Cost of Inventory—This value, $20, is the average manufacturing cost of one pair of Green Jeans.
- Sales Price per unit—This value, $30, is the average sales price to the wholesaler or retailer for one pair of Green Jeans.

Sales and Operations Plan Section

This section is the heart of your decision model. All Sales and Operations plans share a similar structure: they contain a section for historical data (in this case, the previous three months) that includes both the plan figures and the actual results for each month. Next to the historical data is the plan for the next six months. As each month's operations are completed and the actual sales data is collected, the column for that month will move to the History section, and a new month will be added to the plan. As stated before, most Sales and Operations plans cover 12 months, but Green Jeans' plans are for six months.

To avoid excessive data entry in the historical section, you should download the skeleton file for this case. If you choose to set up the model yourself, refer to Figure 9-2. A description of each line item follows the figure. Gray cells contain formulas, and the yellow cells are the Changing Cells for this model.

NOTE—DEALING WITH LARGE NUMBERS

The numbers in Sales and Operations plans are often quite large. To keep column widths manageable, planners frequently express sales and units as multiples of thousands or millions. In this case, you will be given conversion units for the formulas; usually you will divide or multiply by 1,000 to make the conversion.

	A	B	C	D	E	F	G	H	I	J	K
12											
13		Sales and Operations Plan		History					Plan		
14			Oct	Nov	Dec	Jan	Feb	Mar	Apr	May	Jun
15		*Sales*									
16		Forecast (in million $)	9.99	13.11	6.90						
17		(in 1,000 units)	333	437	230	253	280	340	300	393	233
18		Actual (in 1,000 units)	300	400	200						
19											
20		*Operations*									
21		Plan (in 1,000 units)	440	440	220						
22		(in employees)	1000	1100	550	550	550	550	550	550	550
23		Number working days/mo	22	20	20	20	21	23	20	22	22
24		Actual (in 1,000 units)	360	455	300						
25											
26		*Inventory (end of month)*									
27		Plan (in 1,000 units)	100	100	100						
28		(in 1,000 $)	3000	3000	3000						
29		Actual (in 1,000 units)	60	115	215						
30		Plan Days of Supply	6.6	4.6	8.7						

FIGURE 9-2 Sales and Operations Plan section

- History—This section contains values for October, November, and December, as shown in Figure 9-2. The actual end-of-month inventory for December is the only value here that you will use in a formula for the plan section. *If you use the spreadsheet skeleton, these values will already be entered for you.*
- "Actual" (Sales, Operations, and Inventory)—Although these lines have entries in the History section, they will remain blank in the Plan section. Accountants use these cells in the worksheet to enter the Actual data for each month.
- Forecast (in million $)—This line contains formulas for the dollar values of the sales units forecast below. The formula for this value is the sales forecast in units for the month multiplied by the sales price per unit, then divided by 1,000 to convert the units (in thousands) into dollars (in millions).
- Forecast (in 1,000 units)—These values are the sales forecast for Green Jeans, in thousands. Enter the following values in the order shown: 253, 280, 340, 300, 393, and 233.
- Operations Plan (in 1,000 units)—This line contains formulas for the amounts of planned production, in thousands of units. The formula for this value is the number of employees multiplied by Employee Productivity (cell $C$4) multiplied by the number of working days per month, divided by 1,000. The number of employees and working days per month are reported in the next two cells.
- Plan (in employees)—This line is the heart of the Solver model: the Changing Cells. These values are the numbers of employees in the workforce each month. Enter 550 in each cell for now. Place a fill color in this line of cells to indicate that they will be the Changing Cells for Solver. For added emphasis, you can select a thick box border to place around this line.
- Number working days/mo.—These values are the number of days worked each month in the plan. The Green Jeans plant runs five days per week. Enter the following values in the order shown: 20, 21, 23, 20, 22, 22.

- Inventory Plan (in 1,000 units)—This line contains formulas for the number of finished jeans that remain in inventory at the end of each month. January's formula (cell F27) is different from the rest of the months because it will use the Actual ending inventory from December (cell E29). Type the following formula into cell F27: =E29+F21−F17. The formula for February through June (cells G27 through K27) will be the previous month's plan inventory plus the plan production for the current month, minus the forecast sales for that month. If you are confused, the formula for cell G27 is =F27+G21−G17. Next, copy cell G27's formula into cells H27 through K27.

- Plan (in 1,000 $)—This line contains the dollar value of the finished jeans that remain in inventory at the end of each month. The formula for this value is the inventory plan (in thousands of units) from the previous cell multiplied by the Unit Cost of Inventory (cell $C$10 in the Constants section).

- Plan Days of Supply—This line contains the days of finished goods inventory available to supply the forecast sales demand for that period. The formula for this value is the plan inventory in units divided by the sales forecast, multiplied by the number of working days in each month. Because both the plan inventory and the sales forecast are in units of 1,000, the formula does not need a conversion factor. If you are confused, the formula for cell F30 is =(F27/F17)*F23. Next, copy cell F30's formula into cells G30 through K30.

If you wrote your formulas correctly, your Sales and Operations Plan section should have the values shown in Figure 9-3. If you have different numbers, check your work.

	B	C	D	E	F	G	H	I	J	K
12										
13	Sales and Operations Plan		History				Plan			
14		Oct	Nov	Dec	Jan	Feb	Mar	Apr	May	Jun
15	*Sales*									
16	Forecast (in million $)	9.99	13.11	6.90	7.59	8.40	10.20	9.00	11.79	6.99
17	(in 1,000 units)	333	437	230	253	280	340	300	393	233
18	Actual (in 1,000 units)	300	400	200						
19										
20	*Operations*									
21	Plan (in 1,000 units)	440	440	220	220	231	253	220	242	242
22	(in employees)	1000	1100	550	550	550	550	550	550	550
23	Number working days/mo	22	20	20	20	21	23	20	22	22
24	Actual (in 1,000 units)	360	455	300						
25										
26	*Inventory (end of month)*									
27	Plan (in 1,000 units)	100	100	100	182	133	46	-34	-185	-176
28	(in 1,000 $)	3000	3000	3000	3640	2660	920	-680	-3700	-3520
29	Actual (in 1,000 units)	60	115	215						
30	Plan Days of Supply	6.6	4.6	8.7	14	10	3	-2	-10	-17

FIGURE 9-3 Sales and Operations Plan section with the formulas correctly entered

Hires and Layoffs and Costs Sections

These sections contain calculations for the number of hires and layoffs in your workforce for each month in the plan, and calculations of the hiring costs, layoff costs, and inventory carrying costs for each month. These sections also contain the total plan cost, which will be the Optimization Cell that you want to minimize. If you downloaded the skeleton worksheet, you will only have to enter the formulas for the cells in these sections. See Figure 9-4 and the following list for the formulas you will enter.

	B	C	D	E	F	G	H	I	J	K	L
31											
32				**Hires and Layoffs**							
33	Color Legend:			No. of:	Jan	Feb	Mar	Apr	May	Jun	Totals
34	Optimization Cell--Combined Cost			Hires							
35	Changing Cells--Manning Level			Layoffs							
36	Cells with Formulas										
37				**Costs**							
38				Cost of:	Jan	Feb	Mar	Apr	May	Jun	Totals
39				Hires							
40				Layoffs							
41				Inventory							
42											

FIGURE 9-4 The Hires and Layoffs and Costs sections with a color legend for the cells

- Color Legend—This section contains a legend that explains the purpose of the fill colors in the worksheet.
- Hires—This line contains the number of new employees hired each month. If the number of employees in the plan for the current month is *greater than* the number of employees in the plan the previous month, then the number of new hires equals the number of employees in the plan for the current month *minus* the number of employees in the plan the previous month. If the number of employees in the plan for the current month is *less than or equal to* the number of employees the previous month, then the number of new hires is zero. If you need a refresher in how to convert IF statements into formulas, see Tutorial C.
- Layoffs—This line contains the number of employee layoffs each month. If the number of employees in the plan for a month is *less than* the number of employees in the plan the previous month, then the number of layoffs equals the number of employees in the plan for the previous month *minus* the number of employees in the plan for the current month. If the number of employees in the plan for the previous month is *less than or equal to* the number of employees for the current month, then the number of layoffs is zero.
- Totals—Cells L34 and L35 contain the total numbers of hires and layoffs. Enter formulas for the sums of monthly hires and layoffs for the six months into cells L34 and L35, respectively.
- Cost of Hires—This value is the number of hires from the Hires and Layoffs section multiplied by the Hiring Cost per worker from the Constants section (cell $C$5).
- Cost of Layoffs—This value is the number of layoffs from the Hires and Layoffs section multiplied by the Layoff Cost per worker from the Constants section (cell $C$6).
- Cost of Inventory—This value is the Plan (in 1000 $) from the Inventory portion of the Sales and Operations Plan section multiplied by the Inventory Carrying Cost from the Constants section (cell $C$7) multiplied by 1,000. If you are confused about this formula, enter the following in cell F41: =F28*1000*$C$7. Copy this formula to cells G41 through K41.
- Totals—Cells L39, L40, and L41 contain the total costs of hiring, layoffs, and inventory (carrying cost). Enter formulas for totaling each of these three costs for the six months in cells L39, L40, and L41, respectively. Cell L42 is the total of cells L39 through L41, and is the Optimization Cell for your decision model. Enter the appropriate formula to sum your costs in this cell.

If you wrote the formulas correctly, your last two sections will look like those in Figure 9-6.

Because you have the same employment level (550) set up from January through June, the Hires and Layoffs section should contain all zeros. The Inventory costs decline steadily by month and actually become negative starting in April, because 550 employees cannot produce enough jeans per month to keep up with the sales forecast, given the starting inventory. This is acceptable because you have not yet specified parameters for Solver to determine your plan.

You have finished the initial setup of your decision model. Right-click the sheet tab at the bottom of the worksheet and rename it **Green Jeans—Starting**. You will copy this worksheet to build each of your Solver solutions. If you have not already saved your workbook, click File, click Save, and then name the file **Green Jeans SOP (*your name*).xlsx.** Include the first initial of your first name and then your last name. Including your name in the filename makes it easier for your instructor to identify the workbook.

Your completed worksheet should look like Figures 9-5 and 9-6.

	B	C	D	E	F	G	H	I	J	K
1	**Green Jeans, LLC Sales and Operations Plan Jan-Jun 2012**									
2										
3	Constants									
4	Employee Productivity	20	units/worker-day							
5	Hiring Cost	$200	per worker							
6	Layoff Cost	$500	per worker							
7	Inventory Carrying Cost	2%	per month applied to ending monthly inventory							
8	Minimum Inventory Level	5	days of supply							
9	Beginning labor force	550	workers							
10	Unit Cost of Inventory	$20								
11	MSRP--unit of Sales	$30								
12										
13	Sales and Operations Plan		History				Plan			
14		Oct	Nov	Dec	Jan	Feb	Mar	Apr	May	Jun
15	*Sales*									
16	Forecast (in million $)	9.99	13.11	6.90	7.59	8.40	10.20	9.00	11.79	6.99
17	(in 1,000 units)	333	437	230	253	280	340	300	393	233
18	Actual (in 1,000 units)	300	400	200						
19										
20	*Operations*									
21	Plan (in 1,000 units)	440	440	220	220	231	253	220	242	242
22	(in employees)	1000	1100	550	550	550	550	550	550	550
23	Number working days/mo	22	20	20	20	21	23	20	22	22
24	Actual (in 1,000 units)	360	455	300						
25										
26	*Inventory (end of month)*									
27	Plan (in 1,000 units)	100	100	100	182	133	46	-34	-185	-176
28	(in 1,000 $)	3000	3000	3000	3640	2660	920	-680	-3700	-3520
29	Actual (in 1,000 units)	60	115	215						
30	Plan Days of Supply	6.6	4.6	8.7	14	10	3	-2	-10	-17

FIGURE 9-5 Completed Constants and Sales and Operations Plan sections

	B	C	D	E	F	G	H	I	J	K	L
31											
32				Hires and Layoffs							
33	Color Legend:			No. of:	Jan	Feb	Mar	Apr	May	Jun	Totals
34	Optimization Cell--Combined Cost			Hires	0	0	0	0	0	0	0
35	Changing Cells--Manning Level			Layoffs	0	0	0	0	0	0	0
36	Cells with Formulas										
37				Costs							
38				Cost of:	Jan	Feb	Mar	Apr	May	Jun	Totals
39				Hires	$0	$0	$0	$0	$0	$0	$0
40				Layoffs	$0	$0	$0	$0	$0	$0	$0
41				Inventory	$72,800	$53,200	$18,400	-$13,600	-$74,000	-$70,400	($13,600)
42											($13,600)

FIGURE 9-6 Completed Hires and Layoffs and Costs sections

Setting up and Running Solver for the Chase Plan

The first plan you will examine is George Detweiler's Chase proposal. In this plan, you will set up the constraints so that the monthly units produced are at least equal to the sales forecast. Copy your Green Jeans—Starting worksheet to a new worksheet, and rename it **Green Jeans—Chase**.

Before using the Solver Parameters window, you should consider jotting down the parameters you must define and their cell addresses. Here is a suggested list:

- The cell you want to optimize (total cost—cell L42) and whether you want to minimize or maximize it
- The cells you want Solver to manipulate to obtain the optimal solution: Plan (in employees), in cells F22 to K22
- The constraints you will define:
    - Each Changing Cell must be an integer greater than or equal to zero—you cannot hire half an employee.
    - You must keep at least a five-day supply of inventory in every month. In other words, each cell in the Plan Days of Supply row must contain a value of 5 or more.
    - For the Chase plan, your units of production each month must be greater than or equal to the forecast units. For example, the value in cell F21 must be greater than or equal to the value in cell F17 for January. You must repeat this constraint for each month of the plan.

Normally, you would choose Simplex LP for the solving method, but in this case the model is not linear, so Solver would not provide a solution. If you try using Simplex LP, you should receive the error message shown in Figure 9-7.

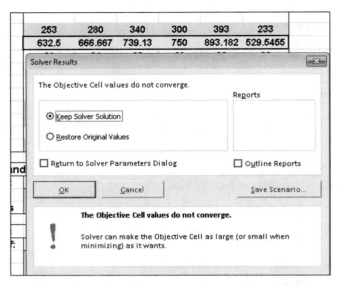

FIGURE 9-7   Error message when trying to use the Simplex LP method

The Simplex LP method will not work for a couple of reasons. First, you did not define an upper bound for your monthly production; you defined the constraint as greater than or equal to the sales forecast. To see what happens when you define an upper bound, add a "less than or equal to" constraint for each month's production in the Solver Parameters window. To open this window, click the Data tab and click Solver in the Analysis group. For example, add $F$21 <= $F$17+10 for January. Repeat the constraint for each month's production. If you run Solver again using the Simplex LP method, you will receive the error message shown in Figure 9-8.

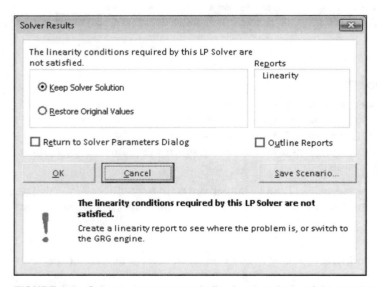

FIGURE 9-8   Solver error message indicating that the model is nonlinear

Solver discovered that some of the equations in your model are nonlinear. Remember the IF statements in the formulas for Hires and Layoffs? Luckily, you can use other solving methods. Follow the advice shown in the Solver Results window, and change your solving method to GRG Nonlinear in the Solver Parameters window (see Figure 9-9). If you added "less than or equal to" constraints for January and the other months, delete the constraints from the Solver Parameters window. However, production must be greater than or equal to the forecast for each month, so leave in those constraints.

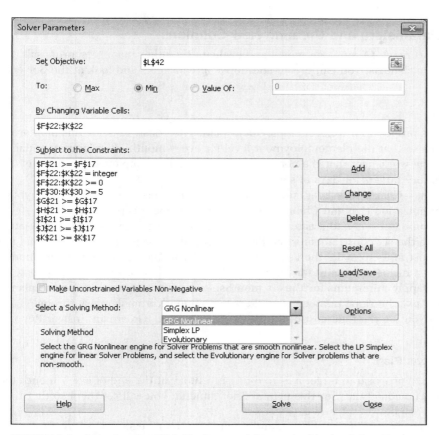

FIGURE 9-9 Selecting the GRG Nonlinear solving method in the Solver Parameters window

Run Solver again. This time you should get a solution that satisfies all your constraints. Click Keep Solver Solution and then create an Answer Report by clicking Answer in the Reports pane of the Solver Results window (see Figure 9-10). Click OK to close the Solver Results window.

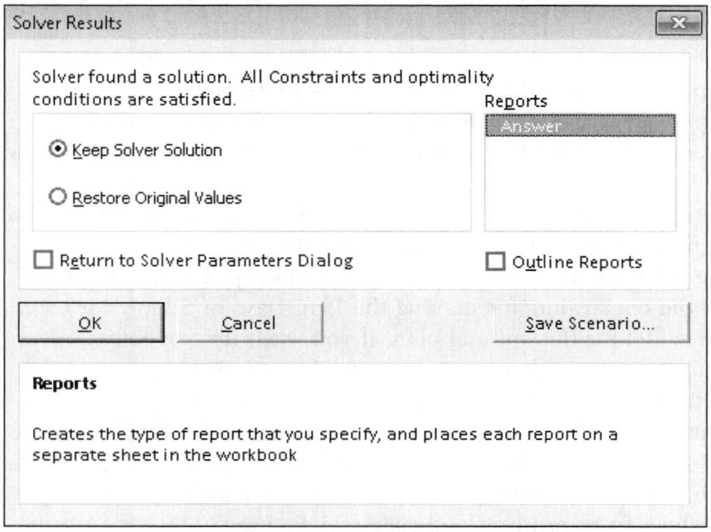

FIGURE 9-10 Creating an Answer Report

You now have the cost data for running the Chase plan. Before continuing, look at the employment levels that Solver chose for optimizing the total cost. Your plan probably shows steady hiring for the first four or five months to meet the sales forecast. By changing the Solver constraints, you can also model a Level plan, a Logistics Target plan, and your own optimized plan.

Assignment 1B: Changing Parameters and Creating New Solutions

Using the model you created in Assignment 1A, you can model and evaluate the other managers' proposals. Now that you know how to copy spreadsheets, you can create additional spreadsheets and look at the Solver solutions for each plan.

Level Sales and Operations Plan

Next, you will evaluate Betty Miller's proposal for a Level employment plan. The constraints for this plan are simple. You will set your constraints so that the plan employment level for every month after January equals January. You know from your original worksheet that the company will have to hire employees to meet its sales demands and minimum inventory targets for days of supply.

Make a copy of your Green Jeans—Chase worksheet, and rename it **Green Jeans—Level Plan**. Open your Solver Parameters window; the Optimization Cell and Changing Cells stay the same. Keep your original constraints for the Changing Cells to be integers and greater than zero, and keep the Plan Days of Supply at greater than or equal to 5, but delete the six constraints for the Production cells to be greater than or equal to the forecast. Next, you must add constraints to make the February, March, April, May, and June employment levels equal to January's. Do not set a hiring constraint for January; Solver must be allowed to "hire" people in January to satisfy the inventory supply minimums for the six months. Run your Solver model for the plan; if you set it up correctly, Solver will use the same number of workers every month from January through June. You might be surprised when you see the total cost of the Level plan proposal. Create an Answer Report for this solution.

5-6 Day Target Sales and Operations Plan

Next, you will evaluate Pedro Sanchez' proposal to target a 5- to 6-day inventory at the end of every month to minimize inventory carrying costs. The constraints for this plan are not difficult; you will set them so that the Plan Days of Supply for each month are between 5 and 6.

Make a copy of your Green Jeans—Level Plan worksheet, and rename it **Green Jeans—5-6 Day Target**. Open your Solver Parameters window; the Optimization Cell and Changing Cells stay the same. Keep your original constraints for the Changing Cells to be integers and greater than zero, and keep the Plan Days of Supply at greater than or equal to 5, but delete the constraints for making all the monthly employment levels equal to January's. Add constraints to make the Plan Days of Supply for each month less than or equal to 6. Run your Solver model for the plan; if you set it up correctly, Solver will keep the Plan Days of Supply between 5 and 6 days for every month from January through June. Create an Answer Report for this solution.

Optimal Sales and Operations Plan

Finally, you will create what you think is the optimal Sales and Operations plan for Green Jeans. The only constraint you think you have to satisfy is the minimum five days of supply for the inventory. You will give Solver complete freedom to hire or fire employees as necessary to minimize the total cost. The constraints for this plan are the easiest to define.

Make a copy of your Green Jeans—5-6 Day Target worksheet, and rename it **Green Jeans—Optimal Plan**. Open your Solver Parameters window; the Optimization Cell and Changing Cells stay the same. Keep your original constraints for the Changing Cells to be integers and greater than zero, and keep the Plan Days of Supply at greater than or equal to 5, but delete the constraints for making the Plan Days of Supply for each month less than or equal to 6. Run your Solver model for the optimal plan. If you set it up correctly, Solver will probably give you the least expensive solution of all the plans. You may also be surprised by its recommended employment levels in the Changing Cells. Create an Answer Report for this solution.

When you have finished all four plans, print a copy of all the worksheets and Answer Reports. Open the Page Setup dialog box by clicking the File tab on the Ribbon, then click Print in the left pane and click Page Setup, the bottom item in the middle pane. The Page Setup dialog box appears. Select Landscape as the page layout for the worksheets, and make each worksheet fit to print on one page (see Figure 9-11). The data in the worksheet should be large enough to read easily.

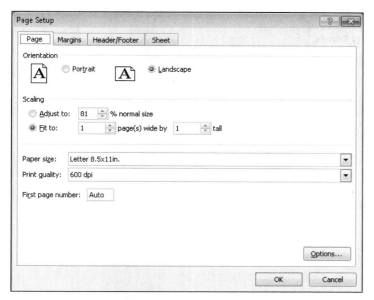

FIGURE 9-11   The Page Setup dialog box

ASSIGNMENT 2: USING THE WORKBOOK FOR DECISION SUPPORT

You have built a series of worksheets to capture each manager's proposal for the Sales and Operations plan, along with your own optimized plan. You will now complete the case by using your solutions and Answer Reports to gather the data needed to make the hiring decisions, and by documenting your recommendations in a memorandum.

Assignment 2A: Using Your Workbook to Gather Data

Use your consolidated workbook and the printouts you created in Assignment 1B to develop a summary table of results that you will place in your memo. The easiest way to make this table is to create a new worksheet in your workbook, and then copy and paste the table into your memo. The table format is shown in Figure 9-12.

Key Performance Indicators	Chase Plan	Level Plan	5-6 Day Target	Optimal Plan
Green Jeans Sales and Operations Plan--Summary of Results				
Manning--January				
Manning--June				
Total Hires				
Total Layoffs				
Hiring Cost--Jan to Jun				
Layoff Cost--Jan to Jun				
Inventory Cost--Jan to Jun				
Total Cost				

FIGURE 9-12   Format of table to be used in memo

A 3-D column chart of the total cost will add visual impact to your presentation. You can insert the chart in the same worksheet as the table, as shown in Figure 9-13. To start a chart, select your four model headings (cells C4 to F4), and then hold down the Ctrl key to select the four Total Cost cells (cells C12 to F12). Next, click the Insert tab on the Ribbon. In the Charts group, click Column, and then select 3-D Column from the menu.

A 3-D column chart is placed in the worksheet. Click it to select it, and reposition the chart beneath your table. Enter an appropriate title for the chart. You can watch the chart values become columns as you fill in the Total Cost cells. In Figure 9-13, the value in cell C12 is *not* the correct solution for the Chase Plan total cost. It was entered to illustrate the graph as it was being built.

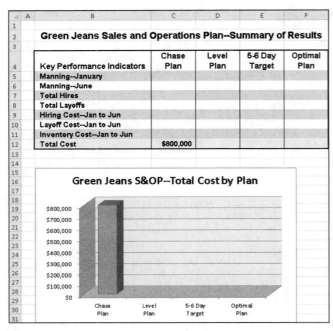

FIGURE 9-13 Building a 3-D column chart while populating your table

Assignment 2B: Documenting Your Recommendations in a Memo

Use Microsoft Word to write a brief memo to the company president, Kathy Green. State the results of your analysis and recommend which Sales and Operations plan you think Green Jeans should adopt. Recall from earlier in the case that Kathy is leaning toward the HR manager's proposal, which would not lay off workers. Does your optimal plan lay off any workers? Your memo should observe the following requirements:

- Set up your memo as described in Tutorial E.
- In the first paragraph, briefly describe the situation and state the purpose of your analysis.
- Next, summarize the results of your analysis and state your recommended action.
- Support your recommendation with your Summary of Results table copied from the Excel workbook. (For a brief description of how to copy and paste Excel objects, see Tutorial C.)
- You should also copy your chart from the Excel workbook into the memo for maximum visual impact.
- In a future case, you might suggest extending the Sales and Operations plan to 12 months instead of six to capture all the seasonal trends in clothing retail. If the Sales Department can provide an accurate 12-month sales forecast, you could recalculate the different proposals for the plan.

ASSIGNMENT 3: GIVING AN ORAL PRESENTATION

Your instructor may request that you give your analysis and recommendations in an oral presentation. If so, assume that company president Kathy Green wants your presentation to last 10 minutes or less. When preparing your presentation, use PowerPoint slides or handouts that you think are appropriate. Tutorial F explains how to prepare and give an effective oral presentation.

DELIVERABLES

Prepare the following deliverables for your instructor:

1. A printout of the memo
2. Printouts of all your worksheets and Answer Reports
3. Your Word document, Excel workbook, and PowerPoint presentation on electronic media or sent to your course site as directed by your instructor

Assemble your paper documentation neatly and staple it together with the memo on top. If you used more than one Excel workbook (.xlsx) file for your case, include a note for your instructor that describes the different files. Good luck!

PART 4

CASE **10**

THE COLLEGE RETURN ON INVESTMENT ANALYSIS

Decision Support Using Excel

PREVIEW

The cost of going to college can be considered an investment. Generally, the cost is justified if a college graduate's future earnings sufficiently exceed the future earnings of a high school graduate. Under what specific circumstances are college costs justified? In this case, you will use Excel to answer that question.

PREPARATION

- Review spreadsheet concepts discussed in class and in your textbook.
- Complete any exercises that your instructor assigns.
- Complete any parts of Tutorials C and D that your instructor assigns, or refer to them as necessary.
- Review file-saving procedures for Windows programs, as discussed in Tutorial C.
- Refer to Tutorial E as necessary.

BACKGROUND

The direct costs of higher education are tuition and fees, room and board, books, and supplies. Also, a college student does not work full time while in school, and the indirect cost of those lost wages can be quite high.

A college graduate usually earns more per year than a high school graduate. Therefore, you can think of the cost of higher education as an investment. Money is paid up front to cover college costs for a few years, but each year after graduation, a college graduate's earnings can be considered a "return" on the investment. Over the course of a career, the total earnings can far exceed the cost of the initial investment, and so most people think that a college degree is a good investment.

However, college costs have increased significantly for many years, in good times and bad. Furthermore, many graduates in recent years have not found jobs that pay significantly more than a high school graduate can earn. With the cost of the investment going up, and the returns going down, this case poses the question: Is a college degree still a good investment? You will develop a spreadsheet model to create "what if" scenarios with the significant variables and attempt to answer the question.

The spreadsheet model will contain the following major elements:

- Data about the college under consideration—This information includes costs for yearly tuition and fees, room and board, and books and supplies, as well as the expected rates of increase for each cost. This data would also include the yearly salary and benefits expected for a graduate of the school. For example, a school might charge $20,000 in yearly tuition and fees, $12,000 for room and board, and $3,000 in books and supplies. The expected annual rates of increase could be 10%, 8%, and 7%. A student might want to get an Accounting degree, and the school's Accounting graduates might have an average starting salary of $50,000 per year.
- Data about the student—This information includes the number of years the student expects to need to complete the degree, the part of the country where the student will seek employment, and the annual salary the person would expect to earn as a high school graduate. For example,

the student might think she needs five years to complete a degree, and that she might choose to work in the U.S. Southwest, where pay rate increases have recently been better than in the Northeast.

- Low-risk investment rate—Investment cash flows are negative at first because of the college investment costs, but they are followed by positive cash flows due to yearly returns. To compute the current value, you must discount yearly cash flows back to the present using an appropriate rate of interest. For example, say that you are promised $106 one year from now, and that you can reliably earn 6% on your investments. The current value of the $106 to be received one year from now is $100. In the college investment model, money is invested for four or more years, followed by returns for a 30-year working career or more. For most people, an appropriate low-risk interest rate would be the rate on long-term U.S. bonds. This rate typically has been about 4%, although in recent years it has been lower.

- Data about the region—Given the region the student wants to work in, the model must use the proper rate of salary increase both for college and high school graduates.

- Cost of attending college—Given the costs of the target school, the related rates of increase, and the expected number of years needed to complete the degree, the model would compute the yearly total cost of attendance. These values are the investment cash flows.

- Cash flows in working career—Compute the yearly salaries of a college graduate and a high school graduate under the same circumstances, and compute the yearly differences. The differences are the investment cash inflows.

- Computed net present value—Given the investment costs, investment inflows, and interest rate, compute the present value. A positive value indicates that the investment in college was worthwhile. A negative value indicates that the investment was not worthwhile.

ASSIGNMENT 1: CREATING A SPREADSHEET FOR DECISION SUPPORT

In this assignment, you will produce a spreadsheet that models the problem. Then, in Assignment 2, you will write a memorandum that explains your findings. In Assignment 3, you may be asked to prepare an oral presentation of your analysis.

A spreadsheet has been started for you, and is available for you to use. If you want to use the spreadsheet skeleton, go to your data files, select Case 10, and then select **CollegeROI.xlsx**. Your worksheet should have the following sections:

- Inputs
- Summary of Key Results
- Calculations
- Cost of Attending College and Working Career Inflows
- Net Present Value Calculations

A discussion of each section follows.

Inputs Section

Your spreadsheet should have the inputs shown in Figure 10-1. Note that the values shown are illustrative only. An explanation of the line items follows the figure.

	A	B
1	**COLLEGE RETURN ON INVESTMENT ANALYSIS**	
2		
3	**INPUTS**	
4	DATA ABOUT THE COLLEGE:	
5	TUITION AND FEES	$ 20,000
6	ROOM AND BOARD	$ 12,000
7	BOOKS AND SUPPLIES	$ 5,000
8	COST INCREASE: TUITION AND FEES	10%
9	COST INCREASE: ROOM AND BOARD	8%
10	COST INCREASE: BOOKS AND SUPPLIES	7%
11	COLLEGE GRAD SALARY & BENEFITS	$ 50,000
12		
13	DATA ABOUT YOU:	
14	NUMBER OF YEARS IN COLLEGE (4..6)	4
15	WORK REGION (NE,SE,NW,SW,MID)	**NE**
16	HIGH SCHOOL GRAD SALARY & BENEFITS	$ 35,000
17		
18	LOW RISK INVESTMENT RATE (XX)	7%

FIGURE 10-1   Inputs section

- Tuition And Fees—This value is the current year's tuition and fees for the school of interest.
- Room And Board—This value is the current year's cost of room and board at the school of interest. Assume that this cost would be about the same if the student had an apartment.
- Books And Supplies—This value is the expected cost of books and supplies for the current year at the school of interest.
- Cost Increases—These values are the expected rates of cost increases at the school of interest. Most colleges are fairly predictable about these increases. In the example shown, the school is expected to increase tuition and fees 10% each year.
- College Grad Salary & Benefits—This value is the expected average salary and benefits for this year's graduates, for the school and degree of interest.
- Number Of Years In College—This value is the number of years that the student expects to need to graduate with the degree sought. Enter a number between 4 and 6.
- Work Region—This value is the area of the country where the graduate expects to work. You can enter abbreviations for Northeast, Southeast, Southwest, Northwest, and Midwest.
- High School Grad Salary & Benefits—This value is the current starting yearly salary and benefits for a person who went to work out of high school instead of attending college.
- Low Risk Investment Rate—This value is the assumed rate for a safe long-term investment.

You should appropriately format the cells for numbers, currency, text, or percentages, as shown in the example. Your instructor may require you to insert a comment in one or more cells, and/or use conditional formatting in one or more cells. In the example spreadsheet, a comment appears in the Number Of Years In College cell ("Enter a number between 4 and 6"). The presence of the comment is indicated by the diamond in the upper-right corner of the cell; to see the comment, place the mouse cursor over the diamond. To enter a comment, right-click the cell and choose Insert Comment from the menu. To use conditional formatting, click the Home tab, click Conditional Formatting in the Styles group, and then select Highlight Cell Rules from the menu. For example, you can build a rule that shows the Number Of Years In College cell in red if the user enters a value that is greater than 6.

Summary of Key Results Section

Your worksheet should include the key results shown in Figure 10-2. The values are echoed from other locations in your spreadsheet. An explanation of the line items follows the figure.

	A	B	C	D	E	F
20	**SUMMARY OF KEY RESULTS**					
21	NPV		TOTAL COST OF ATTENDING COLLEGE			
22	WAS IT WORTH IT?		TOTAL BENEFIT OF WORKING CAREER			

FIGURE 10-2   Summary of Key Results section

- NPV—This value is the net present value of the investment's cash flows, discounted using the investment rate from the Inputs section.
- Was It Worth It?—The investment is worthwhile if the NPV is positive; otherwise, the investment is not worth it.
- Total Cost Of Attending College—This value is the undiscounted sum of all college costs.
- Total Benefit Of Working Career—The difference between the college graduate's salary and the high school graduate's salary is calculated for each year of a working career. This value is the undiscounted sum of those differences.

Calculations Section

This section calculates the values shown in Figure 10-3. Values are calculated by formula, not hard-coded. The values are then used in other calculations that follow. An explanation of the line items follows the figure.

	A	B
24	**CALCULATIONS**	
25	DATA ABOUT WORK REGION:	
26	HIGH SCHOOL GRAD SALARY INCREASE	
27	COLLEGE GRAD SALARY INCREASE	

FIGURE 10-3 Calculations section

- High School Grad Salary Increase—The expected work region is an input. Expected annual salary and benefit increases for high school graduates are: NE, 1%; SE, 3%; SW, 4%; NW, 1%; MID, 2%.
- College Grad Salary Increase—Expected annual salary and benefit increases for college graduates are: NE, 2%; SE, 4%; SW, 5%; NW, 3%; MID, 4%.

Cost of Attending College and Working Career Inflows Sections

This section is the spreadsheet body, as shown in Figure 10-4. Values shown are for illustrative purposes only. Investment outflows are computed in the Cost of Attending College section, and investment inflows are computed in the Working Career Inflows section. An explanation of the line items follows the figure.

	A	B	C	D	E	F	G	H	I	J
			YEAR 1	YEAR 2	YEAR 3	YEAR 4	YEAR 5	YEAR 6		
30	**COST OF ATTENDING COLLEGE**									
31	TUITION AND FEES		$ 22,000	$ 24,200	$ 26,620	$ 29,282	$ -	$ -		
32	ROOM AND BOARD		12,960	$ 13,997	$ 15,117	$ 16,326	$ -	$ -		
33	BOOKS AND SUPPLIES		5,350	$ 5,725	$ 6,125	$ 6,554	$ -	$ -		
34	HIGH SCHOOL GRAD SALARY FOREGONE		35,350	$ 35,704	$ 36,061	$ 36,421	$ -	$ -		
35	TOTAL		75,660	$ 79,625	$ 83,922	$ 88,583	$ -	$ -		
36									**WORK**	**YEARS**
37	**WORKING CAREER INFLOWS**								**YEAR 1**	**YEAR 2**
38	COLLEGE GRAD SALARY		--	--	--	--	--	--	$ 55,204	$ 56,308
39	HIGH SCHOOL GRAD SALARY		--	--	--	--	--	--	$ 36,785	$ 37,153
40	DIFFERENCE = COLLEGE ADVANTAGE		--	--	--	--	--	--	$ 18,419	$ 19,155

FIGURE 10-4 Cost of Attending College and Working Career Inflows sections

- Tuition And Fees—Assume that the student is a high school senior and will attend college in the following year. Thus, the Year 1 tuition and fees is the current year amount multiplied by (1 + the expected percentage increase). The Year 2 amount is (1 + the expected percentage increase) multiplied by the Year 1 amount, and so on. In the example shown, the student expects to be in college for four years, so the dashes represent zeroes in years 5 and 6. Positive values are shown for years 5 and 6 if the student enters 5 or 6 for the number of years in college. Thus, an IF statement is needed in the Year 5 and Year 6 cells.
- Room And Board—The logic used for Tuition and Fees is applied here for room and board. Again, IF statements are needed for the Year 5 and Year 6 cells.

- Books And Supplies—The logic used for Tuition and Fees is applied here for books and supplies. Again, IF statements are needed for the Year 5 and Year 6 cells.
- High School Grad Salary Foregone—Students in college are not earning a salary as working high school graduates. Thus, the salary foregone is a cost of attending college. Again, assume that the student is a high school senior and will attend college in the following year. Thus, the salary foregone in Year 1 is the current high school graduate's salary multiplied by (1 + the high school graduate's expected salary percentage increase for the work region). The Year 2 amount is (1 + the expected percentage increase) multiplied by the Year 1 amount, and so on. Positive values are shown for years 5 and 6 if the student enters 5 or 6 for the number of years in college. Thus, an IF statement is needed in the Year 5 and Year 6 cells.
- Total—The total for a year is the sum of all the costs in the year.
- College Grad Salary—The dashes in the initial six cells indicate that they will not contain formulas or values. Assume that the working career will last 30 years after graduation; only two years of that period are shown in Figure 10-4. Year 1 refers to the first year of work, Year 2 to the second year, and so on. Values in years 2 through 30 are equal to the prior year's value multiplied by (1 + the percentage increase in the College Grad Salary & Benefits from the Inputs section). Assume that the student is a high school senior and will attend school in the following year. Thus, the working Year 1 value is the College Grad Salary & Benefits for the current year multiplied by (1 + the expected College Grad Salary Increase from the Calculations section), raised to the proper exponent. For example, if the student will be in school for four years, the current salary is $50,000, and the expected rate of increase in salary is .03, the working Year 1 value would equal $50,000 * ((1 + .03)^5).
- High School Grad Salary—The same logic used to compute the College Grad Salary is applied to calculate the High School Grad Salary.
- Difference = College Advantage—This value is the difference between College Grad Salary and High School Grad Salary. Presumably, the college graduate value exceeds the high school graduate value, so the difference is a positive value that indicates a return on the investment.

Net Present Value Calculations Section

This section starts with the row labeled "NET," as shown in Figure 10-5. The values shown are for illustrative purposes only.

	A	B	C	D	E	F	G	H	I	J
29					COLLEGE	YEARS				
30	COST OF ATTENDING COLLEGE		YEAR 1	YEAR 2	YEAR 3	YEAR 4	YEAR 5	YEAR 6		
31	TUITION AND FEES		$ 22,000	$ 24,200	$ 26,620	$ 29,282	$ 32,210	$ -		
32	ROOM AND BOARD		$ 12,960	$ 13,997	$ 15,117	$ 16,326	$ 17,632	$ -		
33	BOOKS AND SUPPLIES		$ 5,350	$ 5,725	$ 6,125	$ 6,554	$ 7,013	$ -		
34	HIGH SCHOOL GRAD SALARY FOREGONE		$ 35,350	$ 35,704	$ 36,061	$ 36,421	$ 36,785	$ -		
35	TOTAL		$ 75,660	$ 79,625	$ 83,922	$ 88,583	$ 93,640	$ -		
36									WORK	YEARS
37	WORKING CAREER INFLOWS								YEAR 1	YEAR 2
38	COLLEGE GRAD SALARY		--	--	--	--	--	--	$ 56,308	$ 57,434
39	HIGH SCHOOL GRAD SALARY		--	--	--	--	--	--	$ 37,153	$ 37,525
40	DIFFERENCE = COLLEGE ADVANTAGE		--	--	--	--	--	--	$ 19,155	$ 19,910
41										
42	NET		$ (75,660)	$ (79,625)	$ (83,922)	$ (88,583)	$ (93,640)	$ -	$ 19,155	$ 19,910
43	NPV AT LOW RISK INVESTMENT RATE		$ (94,553)							
44	TOTAL COST OF ATTENDING COLLEGE		$ (421,430)							
45	TOTAL BENEFIT OF WORKING CAREER		$ 991,942							
46	WAS COLLEGE WORTH IT?		NO							

FIGURE 10-5 Net Present Value Calculations section

- Net—This row captures the series of investment and return cash flows. Later, the NPV calculation needs a series that starts with negative investment values, so the college year investment values are multiplied by –1 in this row. The work year values are simply echoed from the preceding Difference row.
- NPV At Low Risk Investment Rate—According to Excel's Help, the NPV function calculates the present value of an investment using a discount rate and a series of future investments (negative

numbers) and returned income (positive values). The syntax is =NPV(rate, Value1, Value2, ..., Value N). The rate is an annual rate, and can be a cell reference; in the model, it is an input cell. Value1 ... Value N need not refer to contiguous cell references. Here, the investment values and return values are in the Net row. This row will contain values of zero if the student completes college in less than six years, as illustrated in Figure 10-5. You cannot refer to cells that contain values of zero or to empty cells when calculating the NPV. Thus, you need to create an IF statement and key it to the number of years in college. For example, if the student graduates in four years, the list of cells would refer to C42:F42, I42:AL42. Note that if the NPV is positive, the investment was a good one at the investment rate specified; the discounted benefits received exceeded the discounted costs incurred. But, if the NPV is negative, the investment was not a good one at the investment rate specified; the discounted benefits received were less than the discounted costs incurred.

- Total Cost Of Attending College—This value is the undiscounted sum of the costs of attending college. Note that the SUM function ignores empty or zero-valued cells.
- Total Benefit Of Working Career—This value is the undiscounted sum of the Difference cells in the 30 working years.
- Was College Worth It?—An IF statement is needed. You will see an output of "YES" if the NPV is positive, and "NO" if the NPV is negative. In the example in Figure 10-5, the investment was not worthwhile.

ASSIGNMENT 2: USING THE SPREADSHEET FOR DECISION SUPPORT

You will now complete the case by using the spreadsheet model to gather data needed to answer four questions that a student might have about the value of a college degree. Next, you will document your findings in a memorandum. If your instructor requires it, you will also give an oral presentation.

Assignment 2A: Using the Spreadsheet to Gather Data

You have built the spreadsheet to develop "what if" scenarios using the model's input values. The inputs represent the logic of a question, and the outputs provide information needed to answer the question.

This section provides guidelines for the four questions that your spreadsheet can answer for you.

Question 1: What is the financial impact of taking more than four years to complete the college degree? In other words, is there really a serious financial penalty for staying in school for an extra year or two? The inputs for answering this question are shown in Figure 10-6.

Data About the College:	
Tuition and Fees	20,000
Room and Board	12,000
Books and Supplies	5,000
Cost Increase: Tuition and Fees	10%
Cost Increase: Room and Board	8%
Cost Increase: Books and Supplies	7%
College Grad Salary & Benefits	50,000
Data About You:	
Number of Years in College (4, 5, or 6)	4, 5, 6
Work Region (NE, SE, NW, SW, MID)	NE
High School Grad Salary & Benefits	35,000
Low Risk Investment Rate (XX)	1%, 4%, 7%

FIGURE 10-6 Question 1 input data

Enter the inputs. The input data that changes is the number of years in college and the investment rate. Enter the data, observe the NPV value in the Key Results area, and then manually record the NPV value in a data collection area at the top of the Inputs area. You will record nine data points. Figure 10-7 shows an example of the finished spreadsheet.

	A	B	C	D	E	F	G
1	**COLLEGE RETURN ON INVESTMENT ANALYSIS**						
2							
3	**INPUTS**			**YEARS**	**ONE %**	**FOUR %**	**SEVEN %**
4	DATA ABOUT THE COLLEGE:			FOUR	$ 455	$ 33	$ (55)
5	TUITION AND FEES	$ 20,000		FIVE	$ 280	$ 11	$ (195)
6	ROOM AND BOARD	$ 12,000		SIX	$ 100	$ (33)	$ (260)
7	BOOKS AND SUPPLIES	$ 5,000					
8	COST INCREASE: TUITION AND FEES	10%					
9	COST INCREASE: ROOM AND BOARD	8%					
10	COST INCREASE: BOOKS AND SUPPLIES	7%					
11	COLLEGE GRAD SALARY & BENEFITS	$ 50,000					
12							
13	DATA ABOUT YOU:						
14	NUMBER OF YEARS IN COLLEGE (4..6)	4					
15	WORK REGION (NE,SE,NW,SW,MID)	NE					
16	HIGH SCHOOL GRAD SALARY & BENEFITS	$ 35,000					
17							
18	LOW RISK INVESTMENT RATE (XX)	1%					

FIGURE 10-7   Question 1 data

Note that the results shown in Figure 10-7 are for illustration purposes only, and the nine data points are shown in thousands. You can copy the data to a new worksheet for charting purposes, as shown in Figure 10-8. Name the sheet **Years**.

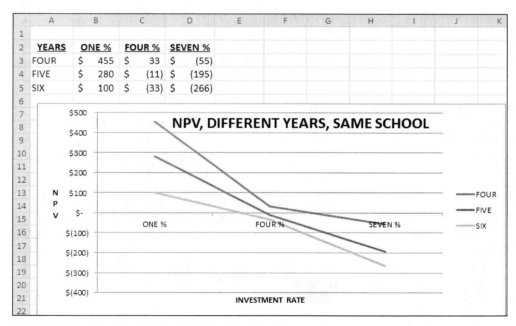

FIGURE 10-8   Question 1 data shown in a line graph

To create an Excel chart, highlight the data set, including the labels. Next, click the Insert tab and then select the type of chart you want in the Charts group. For example, Figure 10-8 shows a Scatter diagram with smooth lines. In the Labels group of the Layout tab, use the Chart Title and Axis Title buttons to create titles and labels.

Make a note of your answer to Question 1 on the face of the spreadsheet or on paper. You will record your answer in a memo later.

At a 4% investment rate, the illustrative data shows that completing the degree in four years is a worthwhile investment because the NPV is positive. If the student stays for a fifth year, the degree is still worth it. However, it is not worthwhile if the student stays for two extra years. At 1%, the investment is worthwhile regardless of whether the student stays for four, five, or six years, but notice how much better the value is if the student graduates in four years rather than six. The two extra years at college amount to a $355,000 penalty! From this data, you would have to conclude that prolonging college for a year or two can result in a significant financial penalty.

Question 2: How do two schools compare in their financial impacts? In this example, assume that the private school charges much more than the state school, but that the target degree and job are the same. The private school's prestige means a somewhat higher starting salary. The inputs for answering the question are shown in Figure 10-9.

Data About the College:	State	Private
Tuition and Fees	25,000	55,000
Room and Board	13,000	30,000
Books and Supplies	3,000	5,000
Cost Increase: Tuition and Fees	11%	10%
Cost Increase: Room and Board	10%	9%
Cost Increase: Books and Supplies	7%	7%
College Grad Salary & Benefits	50,000	53,000
Data About You:		
Number of Years in College (4, 5, or 6)	5	5
Work Region (NE, SE, NW, SW, MID)	SE	SE
High School Grad Salary & Benefits	35,000	35,000
Low Risk Investment Rate (XX)	1%, 4%, 7%	1%, 4%, 7%

FIGURE 10-9   Question 2 input data

Enter the inputs for one school, then vary the interest rate. Observe the NPV values in the Key Results area, and then manually record them in a data collection area at the top of the Inputs area. Change the inputs for the other school, and vary the interest rate. Observe and record the NPV values. You should record six data points. Copy the data to a new worksheet for graphing purposes, and name the sheet **Two Colleges**. Create a chart from the data.

Make a note of your answer to Question 2 on the face of the spreadsheet or on paper. Is one school a much better financial choice than the other? If so, play "what if" with the starting salary of the poorer choice at the 4% level to see where the schools are equal financially. Make a note of the break-even salary levels. You will record your answers in a memo later.

Question 3: Assume that a student is not sure where to work after graduation. The regional choices are NE and SW. How much does the work region matter? The inputs for answering the question are shown in Figure 10-10.

Data About the College:	
Tuition and Fees	30,000
Room and Board	15,000
Books and Supplies	4,000
Cost Increase: Tuition and Fees	10%
Cost Increase: Room and Board	11%
Cost Increase: Books and Supplies	8%
College Grad Salary & Benefits	50,000
Data About You:	
Number of Years in College (4, 5, or 6)	4
Work Region (NE, SE, NW, SW, MID)	NE, SW
High School Grad Salary & Benefits	35,000
Low Risk Investment Rate (XX)	1%, 4%, 7%

FIGURE 10-10   Question 3 input data

Assume that the college and high school graduates' starting salaries are the same in the two regions. Of course, the regions differ in their salary increases. How much does that factor matter? Is one region much better than the other for a lifetime of work?

Enter the inputs for the school, then vary the regions and the interest rates. Observe the NPV values in the Key Results area, and then manually record them in a data collection area at the top of the Inputs area. You should record six data points. Copy the data to a new worksheet for graphing purposes, and name the sheet **Two Regions**. Create a chart from the data.

Make a note of your answer to Question 3 on the face of the spreadsheet or on paper. Is one region a much better financial choice than the other? You will record your answer in a memo later.

Question 4: Consider a student who thinks he will need six years to graduate. The student has three job options out of high school. One option would pay poorly compared with his potential earnings as a college graduate, one option would pay adequately, and one would pay well. Should the student go to college or take one of the jobs out of high school? The inputs for answering the question are shown in Figure 10-11.

Data About the College:	Low pay	Medium pay	High pay
Tuition and Fees	20,000	20,000	20,000
Room and Board	10,000	10,000	10,000
Books and Supplies	3,000	3,000	3,000
Cost Increase: Tuition and Fees	10	10	10
Cost Increase: Room and Board	9	9	9
Cost Increase: Books and Supplies	6	6	6
College Grad Salary & Benefits	50,000	50,000	50,000
Data About You:			
Number of Years in College (4, 5, or 6)	6	6	6
Work Region (NE, SE, NW, SW, MID)	NE	NE	NE
High School Grad Salary & Benefits	33,000	41,000	47,000
Low Risk Investment Rate (XX)	1%, 4%, 7%	1%, 4%, 7%	1%, 4%, 7%

FIGURE 10-11   Question 4 input data

Enter the inputs for Low pay, varying the interest rate as you go. Observe the NPV values in the Key Results area, and then manually record them in a data collection area at the top of the Inputs area. Change the high school graduate salary for Medium pay, vary the interest rate, and record the data. Do the same for High pay. In the end, you should record nine data points. Copy the data to a new worksheet for graphing purposes, and name the sheet **3 HS Jobs**. Create a chart from the data.

Pay close attention to the 4% data values. Does it make sense to take any of the high school graduate jobs if the investment rate is 4%? Make a note of your answer on the face of the spreadsheet or on paper. You will record your answer in a memo later.

When you finish gathering data for the four questions, print the model's worksheet for any set of inputs. Print the chart worksheets if your instructor requires it. Then save the spreadsheet one last time.

Assignment 2B: Documenting Your Findings and Recommendation in a Memo

Now you will document your findings in a memo, which should answer the four questions from the previous section. Use the following tips for guidance in preparing your memo in Microsoft Word:

- Your memo should have proper headings such as Date, To, From, and Subject. You can address the memo to the "prospective college student." You should set up your memo as discussed in Tutorial E.
- Briefly outline the situation. However, you need not provide much background—you can assume that readers are familiar with the situation.
- List and then answer the four questions in the body of the memo.
- What general rule can you cite for the value of a high school degree versus a college education? When is one financially preferable to the other? State your conclusion, referring to your research of the four questions as support.
- Include tables and/or charts to support your claims, as your instructor specifies. Tutorial E explains how to create a table in Microsoft Word.

ASSIGNMENT 3: GIVING AN ORAL PRESENTATION

Your instructor may also request that you summarize your analysis and results in an oral presentation. If so, assume that the student is impressed by your analysis and has asked you to give a presentation to her family to explain the results. Prepare to talk to the group for 10 minutes or less. Use visual aids or handouts that you think are appropriate. Tutorial F provides guidance for how to prepare and give an oral presentation.

DELIVERABLES

Assemble the following deliverables for your instructor:

1. Printout of your memo
2. Spreadsheet printouts
3. Flash drive or CD that contains your Word file and Excel file

Staple the printouts together with the memo on top. If you have more than one .xlsx file on your flash drive or CD, write your instructor a note that identifies your spreadsheet model's .xlsx file.

PART 5

INTEGRATION CASES USING ACCESS AND EXCEL

THE BASEBALL OFFENSIVE PERFORMANCE ANALYSIS

Decision Support with Access and Excel

PREVIEW

At the beginning of this baseball season, a semipro league ordered its home-plate umpires to call a greater percentage of pitches as strikes. This change was expected to reduce offensive performance during the season. Offensive performance data is now available for the season just ended and for the prior season. In this case, you will use Access and Excel to compare offensive performance in the past two baseball seasons and to answer other questions about this season's offensive performance.

PREPARATION

- Review database and spreadsheet concepts discussed in class and in your textbook.
- Complete any exercises that your instructor assigns.
- Complete any parts of Tutorials B, C, and D that your instructor assigns, or refer to them as necessary.
- Review file-saving procedures for Windows programs, as discussed in Tutorial C.
- Refer to Tutorial E as necessary.

BACKGROUND

In baseball, a pitch at which a batter swings and misses is a strike. If the batter does not swing at a pitch, the home-plate umpire must rule it a strike or a ball. By rule, a pitch that passes over the plate at a height between the batter's knees and chest is a strike. A pitch outside this "strike zone" should be called a ball. A batter has three strikes to reach base safely; if the pitcher throws four balls in the same at-bat, the batter is awarded a "walk" to first base.

Several years ago, umpires appeared to change the definition of the strike zone. Pitches thrown over the plate at chest height were being called balls, not strikes. In effect, the upper limit of the strike zone became a point just above the batter's belt line. This change helped hitters in two ways. First, more pitches were called balls, so the number of walks increased. Second, most batters find a high strike harder to hit than a low strike, so batters had a better chance to hit the ball and reach base safely.

With the smaller strike zone, then, more batters reached base via both walks and base hits. The increase in base runners led to more offense—in other words, more runs scored per game. Also, the increased numbers of pitches, walks, and base runners made the average game noticeably longer, to the annoyance of many fans.

At the start of the season that just ended, league officials ordered umpires to call pitches strictly by the rules: pitches that passed over the plate at a height between the batter's knees and chest should be called strikes. This change was expected to speed up the game and to reduce offensive performance. Fewer walks and hits were expected, and therefore fewer runs. More strikeouts were expected. Now that the season is over, league officials want to know if offensive performance actually did decrease significantly. League officials have other questions about offensive performance as well.

You have been hired by league baseball officials to analyze offensive performance in the previous two years. An Access database file named **BaseballAnalysis.accdb** has been provided to assist you. The file contains data for batters' performance during the past two seasons. The tables in the file are discussed next.

Figure 11-1 shows the first few records of player biographical data.

PLAYERNUM	Last Name	First Initial	Team	League	Position	AGE
1	SMITH	S	Cardinals	NL	ThirdBase	30
2	JOHNSON	J	Rangers	AL	CenterField	28
3	WILLIAMS	W	Tigers	AL	LeftField	27
4	JONES	J	Reds	NL	RightField	25
5	BROWN	B	Rockies	NL	CenterField	23

FIGURE 11-1   Player biographical data

Field definitions are as follows:

- PlayerNum—A unique number is assigned to each player. This is the table's primary key field.
- Last Name—The player's last name.
- First Initial—The initial letter of the player's first name.
- Team—The player's team.
- League—Teams are in two leagues: American (AL) and National (NL).
- Position—The player's position in the field.
- Age—The player's age in the season just completed.

The table named ThisYearsStats contains data for how each player performed in the season just ended. Figure 11-2 shows the first few records of the table.

PLAYERNU	PA	AB	R	H	1B	2B	3B	HR	BB	SO	RBI
1	698	587	115	183	101	39	1	42	101	75	117
2	573	517	94	185	111	39	3	32	44	94	102
3	648	547	110	179	93	44	2	40	91	94	125
4	649	547	106	177	99	36	3	39	92	124	115
5	639	586	110	196	120	33	9	34	40	134	119

FIGURE 11-2   Player offensive performance this year

Field definitions are as follows:

- PlayerNum—A unique number is assigned to each player. This is the table's primary key field.
- PA—The number of plate appearances for a player this season; in other words, the number of times a player came to bat.
- AB—The number of at-bats the player had this season. This number is less than plate appearances because walks and some other appearances do not count as an at-bat.
- R—The number of runs the player scored in the season.
- H—The number of hits the player had in the season. A hit occurs when a player hits the ball and reaches base as a direct result. Hits are either singles, doubles, triples, or home runs.
- 1B—The number of singles (one-base hits) the batter had this season.
- 2B—The number of doubles (two-base hits) the batter had this season.
- 3B—The number of triples (three-base hits) the batter had this season.
- HR—The number of home runs the batter had this season.
- BB—The number of times the batter reached first base via a walk.
- SO—The number of times the batter struck out.
- RBI—The number of runs batted in for the player this season. A player is credited with an RBI each time a teammate scores as a direct result of the player's base hit, walk, or other action.

The file has another table called LastYearsStats that contains the same types of offensive performance data as the ThisYearsStats table. Thus, two years of offensive performance data are available. Data in the ThisYearsStats table reflects statistics for a year in which home-plate umpires called strikes by the rulebook. The LastYearsStats table reflects statistics for a year in which umpires used a smaller strike zone.

Some players are in the starting lineup almost every day, and they accumulate hundreds of plate appearances in a season. Other players only appear in games when the starters need a rest. Player data is included in a table only if the player had at least 10 plate appearances in each of the past two seasons.

A database file is available for you to use with the following assignments. To get the file, go to your data files, select Case 11, and then select **BaseballAnalysis.accdb**.

ASSIGNMENT 1: MAKING QUERIES IN ACCESS

In this assignment, you will design and run two queries.

ThisYearsData Query

Create a query named ThisYearsData that computes and shows performance measures for all players who had 100 or more plate appearances this year. Your output should look like that in Figure 11-3; only the first few output records are shown.

Last Name	First Initi	Team	Leagu	Position	AGE	PA	SO	Total Bas	Batting A	Slugging	OPS	OnBaseP
SMITH	S	Cardinals	NL	ThirdBase	30	698	75	350	0.312	0.596	1.009	0.413
JOHNSON	J	Rangers	AL	CenterField	28	573	94	326	0.358	0.631	1.039	0.408
WILLIAMS	W	Tigers	AL	LeftField	27	648	94	347	0.327	0.634	1.058	0.423
JONES	J	Reds	NL	RightField	25	649	124	336	0.324	0.614	1.035	0.421
BROWN	B	Rockies	NL	CenterField	23	639	134	349	0.334	0.596	0.973	0.377

FIGURE 11-3 ThisYearsData query output

The following output values are defined specifically for this query:

- Total Bases—This value is the number of bases accumulated by the player's hits. A single is one base, a double is two bases, a triple is three bases, and a home run counts for four bases.
- Batting Average—This value is the ratio of the player's hits to at-bats.
- Slugging Percentage—This value is the ratio of the player's total bases to at-bats. If two players have the same number of hits and at-bats, they will have the same batting average. However, a player who has more doubles, triples, and home runs will have a higher slugging percentage than a singles hitter.
- OPS—This value is the sum of the player's slugging percentage and on-base percentage. OPS is a combined measure of a player's ability to get on base and hit for power.
- OnBasePercentage—This value is the player's hits plus walks divided by at-bats plus walks; it shows how frequently the player reached base.

Note that ratios are formatted for three decimal places. In Design view, highlight the output column and select Properties from the drop-down menu. Select Standard Format and 3 Decimals.

LastYearsData Query

Create a query named LastYearsData that computes and shows performance measures for all players who had 100 or more plate appearances last year. Your output should look like that in Figure 11-4; only the first few output records are shown.

Last Name	First Initi	Team	League	Position	Age Last Yea	PA	SO	Total Ba	Batting A	Slugging Per	OPS	On Base P
SMITH	S	Cardinals	NL	ThirdBase	29	706	71	357	0.318	0.604	1.021	0.417
JOHNSON	J	Rangers	AL	CenterFiel	27	590	97	344	0.367	0.644	1.062	0.418
WILLIAMS	W	Tigers	AL	LeftField	26	664	90	356	0.332	0.635	1.062	0.427
JONES	J	Reds	NL	RightField	24	687	126	346	0.316	0.598	1.012	0.414
BROWN	B	Rockies	NL	CenterFiel	22	667	135	361	0.333	0.592	0.966	0.374

FIGURE 11-4 LastYearsData query output

Note that Age Last Year is one year less than the player's age this year. Other output values are calculated in the same way as the values in the ThisYearsData query.

When you finish the queries, save and close the **BaseballAnalysis.accdb** file.

ASSIGNMENT 2: USING EXCEL FOR DECISION SUPPORT

In this assignment, you will import your two Access queries into Excel worksheets and then develop information needed to answer league officials' questions about offensive performance.

Importing Queries

Open a new file in Excel and save it as **BaseballAnalysis.xlsx**. Then import the ThisYearsData query output into Excel. To import the data, click the Data tab, select Get External Data, and then select From Access. Specify the Access filename, the query name, and where to place the data in Excel (cell A1 is recommended). Rename the worksheet **ThisYear**.

The data will come into Excel as an Excel data table. Select Total Row in the Table Style Options group to add a Totals row to the bottom of the data table. You will probably need to format the ratio columns for three decimal places. The first few rows of the ThisYearsData query output should look like the data shown in Figure 11-5.

	A	B	C	D	E	F	G	H	I	J	K	L	M
1	Last Name	First Initial	Team	League	Position	AGE	PA	SO	Total Bases	Batting Avg	Slugging	OPS	OnBasePercent
2	SMITH	S	Cardinals	NL	ThirdBase	30	698	75	350	0.312	0.596	1.009	0.413
3	JOHNSON	J	Rangers	AL	CenterField	28	573	94	326	0.358	0.631	1.039	0.408
4	WILLIAMS	W	Tigers	AL	LeftField	27	648	94	347	0.327	0.634	1.058	0.423
5	JONES	J	Reds	NL	RightField	25	649	124	336	0.324	0.614	1.035	0.421

FIGURE 11-5   ThisYearsData query output imported

Repeat the preceding instructions to import the LastYearsData query output using a separate worksheet. Name the worksheet **LastYear**.

Using Data Tables and Pivot Tables to Gather Data

Now use the data tables and pivot tables to gather data needed to answer league officials' questions. (Consult Tutorial E if you need help using data tables and pivot tables.)

League officials want to know if offensive performance decreased this season compared with last season. You can answer this question using data table analysis. Also, league officials have the following additional questions:

- Baseball fans and officials sometimes say that a player's best offensive season occurs when the player is 27 years old. Was this statement more true for this year's players or last year's players? You can answer this question by examining the average OPS for both years using pivot tables.
- Who were the league leaders this year and last year in batting average, OPS, and on-base percentage? You can answer this question using data table analysis.
- Traditional baseball fans consider batting average the best measure of offensive performance. However, many modern baseball analysts say that OPS is a broader and therefore better measure. Which measure is actually better? To shed light on the question, use data table analysis to create two all-star teams. One team contains the players with the highest OPS at each position. The other team contains the players with the highest batting average at each position. If the same players make both teams, you might reasonably conclude that the two measures are equally powerful. But, if the two teams have mostly different players, the two measures must be telling different stories.
- What were team batting averages this year versus last year? Which teams improved their batting average this year? You can answer these questions using pivot table analysis.

Data Table Analyses

Use data tables to analyze three sets of data: (1) year-to-year offensive performance, (2) league leaders this year and last year in batting average, OPS, and on-base percentage, and (3) the all-star team based on OPS versus the all-star team based on batting average. Your analysis will address the following questions.

Did the strike zone change appear to decrease offensive performance this year?

Use the Totals rows in the data tables to determine the average number of plate appearances, strikeouts, total bases, batting average, and OPS for each year. Open a new worksheet and name it **Offensive Comparison**. Copy the Totals rows to the new worksheet. Delete columns that you do not need, and format the numbers appropriately. Enter a note below the data that describes the change in year-to-year offensive performance. The data shown in Figure 11-6 is for illustrative purposes only; your data will look slightly different.

	A	B	C	D	E	F	G
1	This year versus last year						
2		Average	Average	Average	Average		
3		Plate	Strike	Total	Batting	Average	
4		Appearances	Outs	Bases	Average	OPS	
5	This Year	360.1	70.0	140.0	0.275	0.780	
6							
7	Last Year	375.4	60.2	150.1	0.285	0.801	
8							
9	Conclusion: Performance did decline this year, versus last year.						
10	There were fewer plate appearances, total bases, lower batting average						
11	and lower OPS. There were more strike outs.						

FIGURE 11-6   Yearly offensive data comparison

The data indicates that offensive performance declined. If this year's values had been the same or better than last year's, you could have concluded that the strike zone change did not lead to worse offensive performance.

Who were the league leaders this year and last year in batting average, OPS, and on-base percentage?

Sort the relevant data table columns from largest to smallest. Summarize the data in a new worksheet called **League Leaders**. Format values appropriately. The data shown in Figure 11-7 is for illustrative purposes only; your data will look slightly different.

	A	B	C	D	E	F
1	League Leaders					
2		This Year			Last Year	
3		Name	Value		Name	Value
4	Batting Average	Monk	0.36		Geerts	0.374
5	On Base %	Brady	0.433		St Pierre	0.455
6	OPS	Wright	1.111		Wright	1.222

FIGURE 11-7   League-leading performances in each year

Did the OPS and batting average leaders form similar all-star teams this year?

Clicking the arrow in a column heading displays a drop-down menu with a Search option at the bottom. You can use search values to create subsets of the data. In this case, you can use the Position column to show data for players at a particular position. Then you can sort the OPS column to find the best OPS performer at that position during the year. Repeat the procedure in the batting average column for the same position. Use these steps at each remaining position to develop OPS and Batting Average all-star teams for this year. Summarize data in a new worksheet with appropriate formatting. Your data should look like that in Figure 11-8.

	A	B	C	D	E	F	G	H	I
1	All Star Team Selections -- OPS versus Batting Average								
2				MAX					MAX
3	Name	Team	Position	OPS		Name	Team	Position	Batting Avg
4	WILSON	WhiteSox	Catcher	0.977		COOPER	Mariners	Catcher	0.315
5	JOHNSON	Rangers	CenterField	1.039		JOHNSON	Rangers	CenterField	0.358

FIGURE 11-8   OPS and Batting Average all-star teams

In Figure 11-8, data for only two positions are shown. Enter a note below the all-star team data that states your conclusion. Do the two performance measures essentially lead to the same teams or not?

Pivot Table Analyses

Use pivot tables to analyze two sets of data: (1) Player performances at the age of 27 versus those in other years, and (2) team batting averages this year and last year.

Is offensive performance best when the player is 27 years old?

Assume that OPS is the best performance measure to use. Using this year's data, create a pivot table on a separate sheet that shows the average OPS by player age. (For example, show the average OPS for all players who are 25 years old, 26, 27, and so on.) The table should also display a count of players at each age. Player Age should be the row label value. The Average OPS and Number of Players are Value fields. In the pivot table, the Player Age column should have a drop-down arrow; use it to set a Value Filter that shows the top eight OPS values by Player Age. In the pivot table, right-click the top value cell, and then sort from largest to smallest. The player age with the best average OPS will appear at the top row of the pivot table values.

Apply the same procedure for last year's values. Copy the table data and paste it into the sheet that contains this year's pivot table data so you can compare the two years. Below the tables, insert a note that states your conclusion about the 27th year rule—does it apply or not? The data shown in Figure 11-9 is for illustrative purposes only; your data will look slightly different.

	A	B	C	D	E	F	G
1	This Year				Last Year		
2							
3	Player Age	Average OPS	Number of Players		Player Age	Average OPS	Number of Players
4	32	0.758	31		31	0.763	31
5	28	0.736	35		26	0.753	64
6	29	0.734	29		28	0.753	29
7	27	0.722	63		27	0.739	38
8	26	0.720	38		30	0.732	38
9	31	0.717	37		25	0.729	40
10	24	0.712	30		32	0.716	29
11	25	0.708	33		23	0.716	31
12	Grand Total	0.725	296		24	0.707	35
13					Grand Total	0.736	335
14							
15	Conclusion: Age 27 was not the best year, either year.						

FIGURE 11-9 Performance by player age

Note that the same number of players last year were ages 32 and 31, so Excel turned the top eight into the top nine.

What were team batting averages this year versus last year? Which teams improved their batting average this year?

Using separate sheets for each year, create pivot tables that show the composite batting average of each team. In other words, sort the players by team, and then compile one team average from the individual player averages. Sort the data alphabetically by team name. Copy last year's data to the sheet that contains this year's pivot table for comparative purposes. Using Excel formulas, compute the difference in averages for each team. Using an Excel IF() function, show which teams had a better batting average this year than last year. Your results should look like those in Figure 11-10.

	A	B	C	D	E	F	G	H	I
1	This Year's Team Averages			Last Year's Team Averages					
2									
3	Team	Average Batting Avg		Team	Average Batting Avg		Difference		Improved?
4	Angels	0.243		Angels	0.252		-0.009		
5	Astros	0.251		Astros	0.266		-0.014		
6	Athletics	0.253		Athletics	0.258		-0.005		
7	BlueJays	0.252		BlueJays	0.262		-0.010		
8	Braves	0.259		Braves	0.260		-0.001		
9	Brewers	0.263		Brewers	0.258		0.005		Improved

FIGURE 11-10 Team batting averages in each year

Use the Countif() function to compute the number of teams that had an improved batting average, and display that value at the bottom of the Improved? column. (See Tutorial E for guidance on using the Countif() function.)

ASSIGNMENT 3: DOCUMENTING FINDINGS IN A MEMORANDUM

In this assignment, you write a memorandum in Microsoft Word that documents your findings. You should briefly describe the strike zone change. Your memo should then list the league officials' questions and the answers you have developed in Excel.

In your memo, observe the following requirements:

- Your memo should have proper headings such as Date, To, From, and Subject. You can address the memo to the baseball officials who hired you. Set up the memo as discussed in Tutorial E.
- Briefly outline the situation. However, you need not provide much background—you can assume that the readers are familiar with baseball.
- Answer the various questions in the body of the memo.
- Support your claims by showing important results in tables. Tutorial E describes how to create a table in Microsoft Word.

DELIVERABLES

Assemble the following deliverables for your instructor:

1. Printout of your memo
2. Spreadsheet printouts, if required by your instructor
3. Electronic media such as a USB key or CD, which should include your Word file, Access file, and Excel file

Staple the printouts together with the memo on top. If you have more than one .xlsx file or .accdb file on your electronic media, write your instructor a note that identifies the files in this assignment.

THE BREWERY INDEX CALCULATION

Decision Support with Access and Excel

PREVIEW

A regional stock exchange has a number of listed stocks from area breweries. The exchange wants to create an "index" that tracks the performance of the brewery stocks. In this case, you will use Access and Excel to create the index.

PREPARATION

- Review database and spreadsheet concepts discussed in class and in your textbook.
- Complete any exercises that your instructor assigns.
- Complete any parts of Tutorials B, C, and D that your instructor assigns, or refer to them as necessary.
- Review file-saving procedures for Windows programs, as discussed in Tutorial C.
- Refer to Tutorial E as necessary.

BACKGROUND

The New York Stock Exchange (NYSE) is not America's only stock exchange. Other large U.S. cities are home to stock exchanges of their own. Such exchanges generally list stocks of companies that are prominent in the region. Your city is home to a regional stock exchange.

Beer and ale brewing is popular in your part of the country, and many small and medium-sized breweries are based in the region. These companies have listed their common stocks on your city's exchange, which lets investors buy and sell shares in the listed stocks.

Frequently, the most common question of investors is whether stock values have gone up or down. Some investors do not like to search the hundreds of listed stocks to find out. Therefore, the exchange needs an index: a single number that describes the price movement for a group of stocks over a period of time.

Perhaps the best-known index is the NYSE's Dow Jones Industrial index. "The Dow" captures the price movement of shares of 30 large companies listed on the NYSE. The Standard & Poor's 500 index performs the same function for 500 important NYSE companies.

Your city's regional exchange wants to create an index devoted to the price movement of its listed brewery stocks, and you have been asked to help. First you must understand how this kind of index is calculated. For example, consider the stock price data for three companies, as shown in Figure 12-1.

	A	B	C
1	Stock	O/S Shares	Price
2	A	1000	$ 10
3	B	2000	$ 8
4	C	3000	$ 12

FIGURE 12-1 Example stock price data

Company A has 1,000 shares of its stock "outstanding" (issued to investors), and its current price is $10 per share. Company B has 2,000 shares outstanding, with a current price of $8 per share. Company C has 3,000 shares outstanding at a price of $12 per share. One way to compute an index for these three companies would be to average the prices, which would result in an index of 10 (30/3).

Note that such an index would be "unweighted"—in other words, each company would be treated equally in the averaging process. However, the market "capitalizations" (the number of shares multiplied by price per share) are not the same. Company C has $36,000 of stock outstanding, B has $16,000, and A has $10,000.

Stock market indexes generally are weighted by capitalization. The capital weighted index for companies A, B, and C would be calculated as shown in Figure 12-2.

	A	B	C	D
1	Stock	O/S Shares	Price	Product
2	A	1000	$ 10	10000
3	B	2000	$ 8	16000
4	C	3000	$ 12	36000
5		6000		62000
6	Index			10.33

FIGURE 12-2 Capital weighted index calculation

The Product column shows the market value for each company. Total market value ($62,000) divided by total shares (6,000) yields an index of 10.33. The effect of a share price change is shown in Figure 12-3.

	A	B	C	D
1	Stock	O/S Shares	Price	Product
2	A	1000	$ 11	11000
3	B	2000	$ 8	16000
4	C	3000	$ 12	36000
5		6000		63000
6	Index			10.50

FIGURE 12-3 Index change caused by price change

Company A's price rose by a dollar, causing the index to increase slightly. Of course, if the price had declined by a dollar, the index would have declined slightly.

A change in shares outstanding can also change the index value, as shown in Figure 12-4.

	A	B	C	D
1	Stock	O/S Shares	Price	Product
2	A	1000	$ 10	10000
3	B	4000	$ 8	32000
4	C	3000	$ 12	36000
5		8000		78000
6	Index			9.75

FIGURE 12-4 Index change caused by change in shares outstanding

In the figure, company B issued 2,000 more shares, also valued at $8. Company B's relatively lower price is now more heavily weighted, moving the index lower.

In recent years, mutual funds have created funds that mirror well-known stock indexes. For example, some funds track the S&P 500 index. Such funds are a collection of stocks in the same ratios as the index. A company that manages this type of fund will buy and sell shares of a stock as its capitalization changes in the index. Therefore, the value of the fund tracks the value of the index.

Some investors think that the value of the overall market will increase over time, but that it is difficult to predict changes in the value of individual stocks. People who want to invest in the market as a whole are attracted to index-based funds.

If there were a brewery index, the regional exchange managers think that a mutual fund company might create a fund that tracks the index. The exchange managers believe that such a mutual fund would be attractive to investors who think the brewery industry will prosper in the coming years. The new fund would also increase the exchange's business in two ways. First, the fund manager would need to buy and sell stocks of the breweries in the index. Second, the brewery index fund would be listed on the exchange, meaning that investors would buy and sell the fund on the exchange.

Investors often try to predict the price changes of individual stocks. Investors who think that a stock's price will increase might buy more shares. Investors who think a stock's price will decrease might sell shares.

Predictions are sometimes based on the earnings per share ratio (EPS), which is the ratio of a company's net income after taxes to the number of its shares outstanding. For example, if a company's net income is $10 million and it has 1 million shares outstanding, its EPS is 10. Most stock market theorists think that a company's stock price is related to its profitability. Thus, the ability to predict a company's EPS should help predict its stock price.

Some predictions are based on price earnings (PE), which is the ratio of the stock's price on the exchange to the company's earnings per share. For example, if a company has an EPS of 10 and its stock sells for $90 on the exchange, its PE ratio would be 9. The ability to predict the PE ratio for a company's stock is helpful in decisions to buy and sell stocks.

Of course, some investors would try to predict changes in the value of the index as a way of anticipating changes in the value of a brewery index fund. Exchange managers wonder how strongly the value of the brewery index correlates with brewery EPS and PE ratios.

You have been hired by the exchange to calculate an index for listed breweries, and to analyze the correlation of the index to company EPS and PE ratios. An Access database file named **Breweries.accdb** has been provided for your assistance; this file contains eight years of data for each brewery. To get the database, go to your data files, select Case 12, and then select Breweries.accdb.

Figure 12-5 shows the first few records of the FinancialData table.

ID	Company Name	Year	Net Income	Outstanding	Price
1	Good Suds	0	$1,130,481	860,123	$14.46
2	Home Town Brewski	0	$1,122,631	590,007	$15.22
3	Golden Brew	0	$3,720,177	333,581	$100.37
4	Light N Zesty	0	$4,237,203	452,523	$112.36
5	Not 2 Bad Brew	0	$1,489,558	344,795	$47.52
6	Bottled Rapture	0	$7,188,402	655,928	$76.71
7	Plains and Mountains	0	$6,366,973	647,583	$117.98
8	Hill and Dale Malt	0	$5,387,854	540,188	$99.74
9	Dark Brood	0	$5,320,784	667,677	$47.81
10	Tough Love	0	$4,110,954	579,455	$49.66
11	Good Suds	1	$1,164,395	911,730	$12.77
12	Home Town Brewski	1	$1,122,631	631,307	$12.45

FIGURE 12-5 Brewery financial data

The following list describes the columns in the table.

- ID—The number of the record. The primary key field.
- Company Name—The name of the company.
- Year—The year number. Year 0 is the first year. Note that the data type is Text.
- Net Income—The company's net income after taxes in the year.
- Outstanding—The number of shares outstanding during the year.
- Price—The share price of the company's common stock on the exchange at the end of the year.

Note that there are 10 brewing companies. Figure 12-5 shows the first-year records for all companies, and second-year records for two of the companies.

ASSIGNMENT 1: MAKING QUERIES IN ACCESS

In this assignment, you will design and run two queries.

EPS Query

Create a query named EPS that computes and shows earnings per share for each company in each year. Your output should look like that in Figure 12-6, although the figure shows only the first few output records.

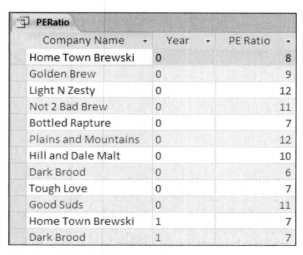

EPS		
Company Name	Year	EPS
Home Town Brewski	0	$1.90
Golden Brew	0	$11.15
Light N Zesty	0	$9.36
Not 2 Bad Brew	0	$4.32
Bottled Rapture	0	$10.96
Plains and Mountains	0	$9.83
Hill and Dale Malt	0	$9.97
Dark Brood	0	$7.97
Tough Love	0	$7.09
Good Suds	0	$1.31
Home Town Brewski	1	$1.78
Dark Brood	1	$9.21

FIGURE 12-6   EPS query output

Note that EPS values are formatted for Currency. If this formatting does not happen automatically, go to Design view, highlight the output column, select Properties in the drop-down menu, and then select Currency Format.

PE Ratio Query

Create a query named PERatio that computes and shows the PE ratio for each company in each year. Your output should look like that in Figure 12-7, although the figure shows only the first few output records.

PERatio		
Company Name	Year	PE Ratio
Home Town Brewski	0	8
Golden Brew	0	9
Light N Zesty	0	12
Not 2 Bad Brew	0	11
Bottled Rapture	0	7
Plains and Mountains	0	12
Hill and Dale Malt	0	10
Dark Brood	0	6
Tough Love	0	7
Good Suds	0	11
Home Town Brewski	1	7
Dark Brood	1	7

FIGURE 12-7   PERatio query output

Note that ratios are formatted for zero decimal places. In Design view, highlight the output column and select Properties in the drop-down menu. Select Standard Number as the Format and zero for Decimal Places. When you finish the queries, save and close the **Breweries.accdb** file.

ASSIGNMENT 2: USING EXCEL FOR DECISION SUPPORT

In this assignment, you will import the Access table that contains your two queries into Excel worksheets and then develop information requested by the exchange management.

Importing and Using Table and Query Data

Open a new file in Excel and save it as **Breweries.xlsx**. To import the FinancialData table into Excel, click the Data tab and then select From Access in the Get External Data group. Specify the Access filename, the table name, and where to place the data in Excel (cell A1 is recommended). Rename the worksheet **FinancialData**.

The data is imported into Excel as an Excel data table. Select Convert to Range in the Tools group to change the data to a range. Format the data for currency and the proper number of decimal places.

By Excel formula, compute the capitalization for each company in each year. The data should look like that in Figure 12-8.

	A	B	C	D	E	F	G
1	ID	Company Name	Year	Net Income	Outstanding	Price	Product
2	1	Good Suds	0	$ 1,130,481	860,123	$ 14.46	$ 12,435,228.27
3	2	Home Town Brewski	0	$ 1,122,631	590,007	$ 15.22	$ 8,981,027.55
4	3	Golden Brew	0	$ 3,720,177	333,581	$ 100.37	$ 33,481,591.69

FIGURE 12-8   Financial data table with capitalization

Note that only the first few rows of the data are shown in Figure 12-8. The capitalization data is shown under a Product heading.

By formula, compute the brewery company index for each year. Summarize the data by year in a convenient spot in the worksheet, and then copy that data to a new worksheet named **Plot Index**. To create a scatter plot of the yearly indexes, highlight the data set, including the labels, then click the Insert tab and select a scatter chart in the Charts group. In the Labels group of the Chart Tools Layout tab, use the Chart Title and Axis Titles buttons to create titles and labels. Your work should look like the illustrative data shown in Figure 12-9.

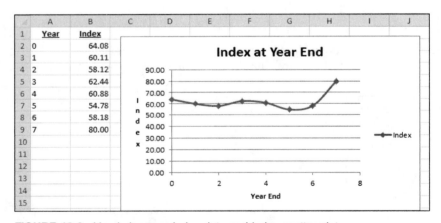

FIGURE 12-9   Yearly brewery index data and index scatter plot

Create a new worksheet and name it **EPS**. Import the EPS query output into the worksheet and format the data appropriately. Compute the unweighted average EPS for each year. (Use the AVERAGE function, and do not weight EPS values by outstanding shares.) Summarize the Average EPS data by year in a convenient spot in the worksheet, and then copy that data to the Plot Index worksheet. Use the directions from the preceding paragraph to create a scatter plot of the yearly values with a proper title and axis labels. Your worksheet should look like the illustrative data shown in Figure 12-10.

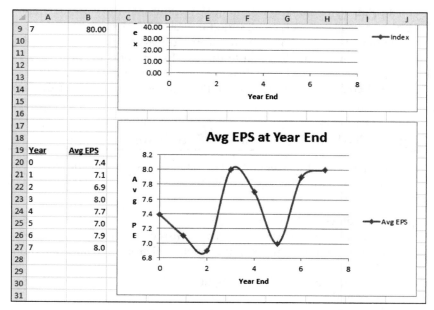

FIGURE 12-10   Yearly EPS data and scatter plot

Use the CORREL function to show the strength of the relationship between the index and average EPS data. Place the formula in a spot that lets you see the relationship easily. Your worksheet should look like Figure 12-11.

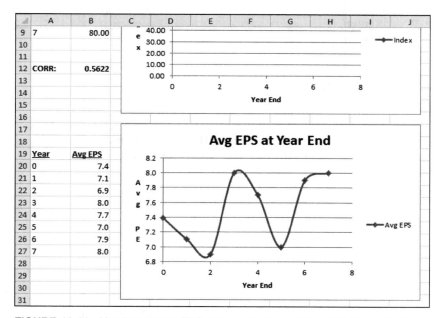

FIGURE 12-11   Yearly average EPS data and index scatter plot with correlation

A correlation value of .56 suggests a weak relationship between the two sets of illustrative data. In fact, the two scatter plots do not look much alike, and so the correlation value confirms the results of the visual inspection. A strong relationship would be suggested by correlation values greater than .70.

Create a new worksheet and name it **PERatio**. Import the PERatio query output into the worksheet and format the data appropriately. Compute the unweighted average PE ratio for each year. (Use the AVERAGE function, and do not weight PE ratio values by outstanding shares.) Summarize the average PE ratio data by year in a convenient spot in the worksheet, and then copy that data to the Plot Index worksheet. Create a scatter plot of the yearly values with a proper title and axis labels. Use the CORREL function to show the strength of the relationship between the index and average PE ratio data. Place the formula in a spot that lets you see the relationship easily.

ASSIGNMENT 3: DOCUMENTING FINDINGS IN A MEMORANDUM

In this assignment, you write a memo in Microsoft Word that documents your findings. You should describe the task that exchange management assigned to you, and then state your findings.

In your memo, observe the following requirements:

- Your memo should have proper headings such as Date, To, From, and Subject. You can address the memo to the exchange management. Set up the memo as discussed in Tutorial E.
- Briefly outline the situation. However, you need not provide much background—you can assume that readers are generally familiar with your task.
- In the body of the memo, describe the index you have computed in Excel. You can present this index numerically, with a chart, or both.
- Then indicate how well the average EPS and average PE values correlate with the brewery index data. Include the correlation data; again, you can support this data numerically, with a chart, or both.
- You can display important results in tables. Tutorial E describes how to create a table in Microsoft Word.

DELIVERABLES

Assemble the following deliverables for your instructor:

1. Printout of your memo
2. Spreadsheet printouts, if required by your instructor
3. Electronic media such as flash drive or CD, which should include your Word file, Access file, and Excel file

Staple the printouts together with the memo on top. If you have more than one .xlsx file or .accdb file on your electronic media, write your instructor a note that identifies this assignment's files.

PART 6

ADVANCED SKILLS USING EXCEL

GUIDANCE FOR EXCEL CASES

The Excel cases in this book require the student to write a memorandum that includes a table. Guidelines for preparing a memo in Microsoft Word and instructions for entering a table in a Word document are provided to begin this tutorial. Also, some of the cases in this book require the use of advanced Excel techniques. Those techniques are explained in this tutorial rather than in the cases themselves:

- Using data tables
- Using pivot tables
- Using built-in functions

You can refer to Sheet 1 of TutEData.xlsx when reading about data tables. Refer to Sheet 2 when reading about pivot tables.

PREPARING A MEMORANDUM IN WORD

A business memo should include proper headings, such as TO, FROM, DATE, and SUBJECT. If you want to use a Word memo template, follow these steps:

1. In Microsoft Word, click File.
2. Click New.
3. Click the Memos button in the Office.com Templates section.
4. Double-click the Contemporary design memo.

The first time you do this, you may need to click Download to install the template.

ENTERING A TABLE INTO A WORD DOCUMENT

Enter a table into a Word document using the following procedure:

1. Click the cursor where you want the table to appear in the document.
2. In the Insert group, select the Table drop-down menu.
3. Select Insert Table.
4. Choose the number of rows and columns.
5. Click OK.

DATA TABLES

An Excel data table is a contiguous range of data that has been designated as a table. Once you make this designation, the table gains certain properties that are useful for data analysis. (Note that in some previous versions of Excel, data tables were called *data lists*.) Suppose you have a list of runners who have completed a race, as shown in Figure E-1.

	A	B	C	D	E	F
1	RUNNER#	LAST	FIRST	AGE	GENDER	TIME (MIN)
2	100	HARRIS	JANE	O	F	70
3	101	HILL	GLENN	Y	M	70
4	102	GARCIA	PEDRO	M	M	85
5	103	HILBERT	DORIS	M	F	90
6	104	DOAKS	SALLY	Y	F	94
7	105	JONES	SUE	Y	F	95
8	106	SMITH	PETE	M	M	100
9	107	DOE	JANE	O	F	100
10	108	BRADY	PETE	O	M	100
11	109	BRADY	JOE	O	M	120
12	110	HEEBER	SALLY	M	F	125
13	111	DOLTZ	HAL	O	M	130
14	112	PEEBLES	AL	Y	M	63

FIGURE E-1   Data table example

To turn the information into a data table (list), highlight the data range, including headings, and select the Insert tab. Then, click Table in the Tables group. The Create Table window appears, as shown in Figure E-2.

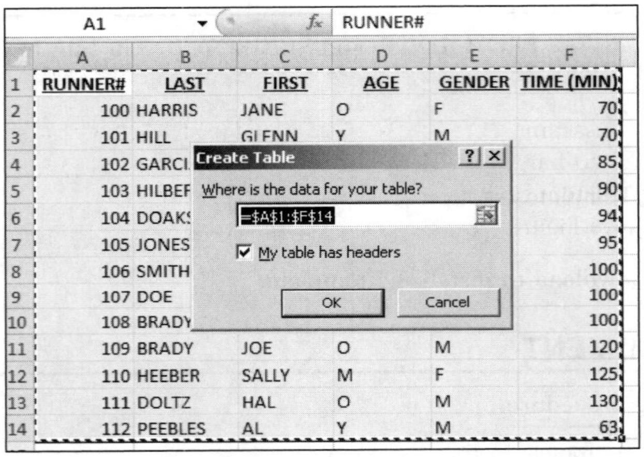

FIGURE E-2   Create Table window

When you click OK, the data range appears as a table. In the Design tab, select Total Row to add a totals row to the data table. You also can select a light style in the Table Styles list to get rid of the contrasting color in the table's rows. Figure E-3 shows the results.

	A	B	C	D	E	F
1	RUNNER#	LAST	FIRST	AGE	GENDER	TIME (MIN)
2	100	HARRIS	JANE	O	F	70
3	101	HILL	GLENN	Y	M	70
4	102	GARCIA	PEDRO	M	M	85
5	103	HILBERT	DORIS	M	F	90
6	104	DOAKS	SALLY	Y	F	94
7	105	JONES	SUE	Y	F	95
8	106	SMITH	PETE	M	M	100
9	107	DOE	JANE	O	F	100
10	108	BRADY	PETE	O	M	100
11	109	BRADY	JOE	O	M	120
12	110	HEEBER	SALLY	M	F	125
13	111	DOLTZ	HAL	O	M	130
14	112	PEEBLES	AL	Y	M	63
15	Total					1242

FIGURE E-3   Data table example

The headings have acquired drop-down menu tabs, as you can see in Figure E-3.

You can sort the data table records by any field. Perhaps you want to sort by times. If so, click the drop-down menu in the TIME (MIN) heading, select Sort, and then select Smallest to Largest. You get the results shown in Figure E-4.

	A	B	C	D	E	F
1	RUNNER ▾	LAST ▾	FIRST ▾	AGE ▾	GENDE ▾	TIME (MIN ▾↑
2	112	PEEBLES	AL	Y	M	63
3	100	HARRIS	JANE	O	F	70
4	101	HILL	GLENN	Y	M	70
5	102	GARCIA	PEDRO	M	M	85
6	103	HILBERT	DORIS	M	F	90
7	104	DOAKS	SALLY	Y	F	94
8	105	JONES	SUE	Y	F	95
9	106	SMITH	PETE	M	M	100
10	107	DOE	JANE	O	F	100
11	108	BRADY	PETE	O	M	100
12	109	BRADY	JOE	O	M	120
13	110	HEEBER	SALLY	M	F	125
14	111	DOLTZ	HAL	O	M	130
15	Total					1242

FIGURE E-4 Sorting list by drop-down menu

You can see that Peebles had the best time and Doltz had the worst time. You also can sort from Largest to Smallest.

In addition, you can sort by more than one criterion. Assume that you want to sort first by gender and then by time (within gender). You first sort from Smallest to Largest in Gender. Then you again click the Gender drop-down tab, select Sort By Color, and select Custom Sort. In the Sort window that appears, click Add Level and choose Time as the next criterion. See Figure E-5.

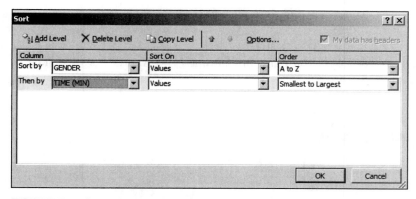

FIGURE E-5 Sorting on multiple criteria

Click OK to get the results shown in Figure E-6.

	A	B	C	D	E	F
1	RUNNER ▾	LAST ▾	FIRST ▾	AGE ▾	GENDE ▾↑	TIME (MIN ▾↑
2	100	HARRIS	JANE	O	F	70
3	103	HILBERT	DORIS	M	F	90
4	104	DOAKS	SALLY	Y	F	94
5	105	JONES	SUE	Y	F	95
6	107	DOE	JANE	O	F	100
7	110	HEEBER	SALLY	M	F	125
8	112	PEEBLES	AL	Y	M	63
9	101	HILL	GLENN	Y	M	70
10	102	GARCIA	PEDRO	M	M	85
11	106	SMITH	PETE	M	M	100
12	108	BRADY	PETE	O	M	100
13	109	BRADY	JOE	O	M	120
14	111	DOLTZ	HAL	O	M	130
15	Total					1242

FIGURE E-6   Sorting by gender and time (within gender)

You can see that Harris had the best female time and that Peebles had the best male time.

Perhaps you want to see the top *n* listings for some attribute; for example, you may want to see the top five runners' times. Select the Time column's drop-down menu, and select Number Filters. From the menu that appears, select Top 10. The Top 10 AutoFilter window appears, as shown in Figure E-7.

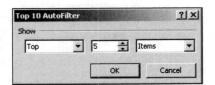

FIGURE E-7   Top 10 AutoFilter window

This window lets you specify the number of values you want. In Figure E-7, five values were specified. Click OK to get the results shown in Figure E-8.

	A	B	C	D	E	F
1	RUNNER ▾	LAST ▾	FIRST ▾	AGE ▾	GENDE ▾↑	TIME (MIN ▾
6	107	DOE	JANE	O	F	100
7	110	HEEBER	SALLY	M	F	125
11	106	SMITH	PETE	M	M	100
12	108	BRADY	PETE	O	M	100
13	109	BRADY	JOE	O	M	120
14	111	DOLTZ	HAL	O	M	130
15	Total					675

FIGURE E-8   Top 5 times

The output contains more than five data records because there are ties at 100 minutes. If you want to see all of the records again, click the Time drop-down menu and select Clear Filter. The full table of data reappears, as shown in Figure E-9.

	A	B	C	D	E	F
1	RUNNER ▾	LAST ▾	FIRST ▾	AGE ▾	GENDE ▾↑	TIME (MIN ▾↑
2	100	HARRIS	JANE	O	F	70
3	103	HILBERT	DORIS	M	F	90
4	104	DOAKS	SALLY	Y	F	94
5	105	JONES	SUE	Y	F	95
6	107	DOE	JANE	O	F	100
7	110	HEEBER	SALLY	M	F	125
8	112	PEEBLES	AL	Y	M	63
9	101	HILL	GLENN	Y	M	70
10	102	GARCIA	PEDRO	M	M	85
11	106	SMITH	PETE	M	M	100
12	108	BRADY	PETE	O	M	100
13	109	BRADY	JOE	O	M	120
14	111	DOLTZ	HAL	O	M	130
15	Total					1242

FIGURE E-9   Restoring all data to screen

Each of the cells in the Total row has a drop-down menu. The menu choices are statistical operations that you can perform on the totals—for example, you can take a sum, take an average, take a minimum or maximum, count the number of records, and so on. Assume that the Time drop-down menu was selected, as shown in Figure E-10. Note that the Sum operator is highlighted by default.

	A	B	C	D	E	F
1	RUNNER ▾	LAST ▾	FIRST ▾	AGE ▾	GENDE ▾↑	TIME (MIN ▾↑
2	100	HARRIS	JANE	O	F	70
3	103	HILBERT	DORIS	M	F	90
4	104	DOAKS	SALLY	Y	F	94
5	105	JONES	SUE	Y	F	95
6	107	DOE	JANE	O	F	100
7	110	HEEBER	SALLY	M	F	125
8	112	PEEBLES	AL	Y	M	63
9	101	HILL	GLENN	Y	M	70
10	102	GARCIA	PEDRO	M	M	85
11	106	SMITH	PETE	M	M	100
12	108	BRADY	PETE	O	M	100
13	109	BRADY	JOE	O	M	120
14	111	DOLTZ	HAL	O	M	130
15	Total					1242 ▾
16						None
17						Average
18						Count
19						Count Numbers
20						Max
21						Min
22						Sum
						StdDev
						Var
						More Functions…

FIGURE E-10   Selecting Time drop-down menu in Total row

By changing from Sum to the Average operator, you find that the average time for all runners was 95.5 minutes, as shown in Figure E-11.

	A	B	C	D	E	F
1	RUNNER ▾	LAST ▾	FIRST ▾	AGE ▾	GENDE ▾↑	TIME (MIN ▾↑
2	100	HARRIS	JANE	O	F	70
3	103	HILBERT	DORIS	M	F	90
4	104	DOAKS	SALLY	Y	F	94
5	105	JONES	SUE	Y	F	95
6	107	DOE	JANE	O	F	100
7	110	HEEBER	SALLY	M	F	125
8	112	PEEBLES	AL	Y	M	63
9	101	HILL	GLENN	Y	M	70
10	102	GARCIA	PEDRO	M	M	85
11	106	SMITH	PETE	M	M	100
12	108	BRADY	PETE	O	M	100
13	109	BRADY	JOE	O	M	120
14	111	DOLTZ	HAL	O	M	130
15	Total					95.53846154 ▾

FIGURE E-11 Average running time shown in Total row

PIVOT TABLES

Suppose you have data for a company's sales transactions by month, by salesperson, and by amount for each product type. You would like to display each salesperson's total sales by type of product sold and by month. You can use a pivot table in Excel to tabulate that summary data. A pivot table is built around one or more dimensions and thus can summarize large amounts of data. Figure E-12 shows total sales cross-tabulated by salesperson and by month.

	A	B	C	D	E
1	Name	Product	January	February	March
2	Jones	Product1	30,000	35,000	40,000
3	Jones	Product2	33,000	34,000	45,000
4	Jones	Product3	24,000	30,000	42,000
5	Smith	Product1	40,000	38,000	36,000
6	Smith	Product2	41,000	37,000	38,000
7	Smith	Product3	39,000	50,000	33,000
8	Bonds	Product1	25,000	26,000	25,000
9	Bonds	Product2	22,000	25,000	24,000
10	Bonds	Product3	19,000	20,000	19,000
11	Ruth	Product1	44,000	42,000	33,000
12	Ruth	Product2	45,000	40,000	30,000
13	Ruth	Product3	50,000	52,000	35,000

FIGURE E-12 Excel spreadsheet data

You can create pivot tables and many other kinds of tables with the Excel PivotTable tool. To create a pivot table from the data in Figure E-12, follow these steps:

1.   Starting in the spreadsheet in Figure E-12, click a cell in the data range, and then click the Insert tab. In the Tables group, choose PivotTable. You see the screen shown in Figure E-13.

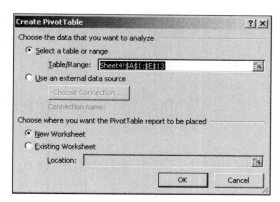

FIGURE E-13   Creating a pivot table

2.   Make sure New Worksheet is checked under "Choose where you want the PivotTable report to be placed." Click OK. The screen shown in Figure E-14 appears. If it does not, right-click in a cell in the pivot table area. Select PivotTable Options from the menu. Click the Display tab and then check the Classic PivotTable layout.

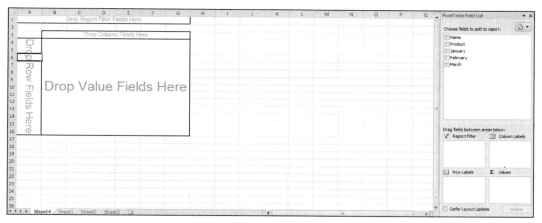

FIGURE E-14   PivotTable design screen

The data range's column headings are shown in the PivotTable Field List on the right side of the screen. From there, you can click and drag column headings into the Row, Column, and Value areas that appear in the spreadsheet.

3.   If you want to see the total sales by product for each salesperson, drag the Name field to the Drop Column Fields Here area in the spreadsheet. You should see the result shown in Figure E-15.

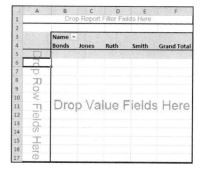

FIGURE E-15   Column fields

4. Next, drag the Product field to the Drop Row Fields Here area. You should see the result shown in Figure E-16.

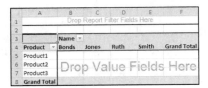

FIGURE E-16   Row fields

5. Finally, drag the month fields (January, February, and March) individually to the Drop Value Fields Here area to produce the finalized pivot table. You should see the result shown in Figure E-17.

Product	Values	Name Bonds	Jones	Ruth	Smith	Grand Total
Product1	Sum of January	25000	30000	44000	40000	139000
	Sum of February	26000	35000	42000	38000	141000
	Sum of March	25000	40000	33000	36000	134000
Product2	Sum of January	22000	33000	45000	41000	141000
	Sum of February	25000	34000	40000	37000	136000
	Sum of March	24000	45000	30000	38000	137000
Product3	Sum of January	19000	24000	50000	39000	132000
	Sum of February	20000	30000	52000	50000	152000
	Sum of March	19000	42000	35000	33000	129000
Total Sum of January		66000	87000	139000	120000	412000
Total Sum of February		71000	99000	134000	125000	429000
Total Sum of March		68000	127000	98000	107000	400000

FIGURE E-17   Data items

By default, Excel adds all of the sales for each salesperson by month for each product. At the bottom of the pivot table, Excel also shows the total sales for each month for all products.

Refer back to Figure E-14 and note the four small panes in the lower-right corner. The Values pane lets you easily change from the default Sum operator to another one (Min, Max, Average, and so on). Click the drop-down arrow, select Value Fields Setting, and then select the desired operator.

BUILT-IN FUNCTIONS

The following functions are referred to in the Excel cases in this text:

• MIN, MAX, AVERAGE, COUNTIF, ROUND, ROUNDUP, and RANDBETWEEN

The syntax of these functions is discussed in this section. The following examples are based on the runner data shown in Figure E-18.

RUNNER#	LAST	FIRST	AGE	GENDER	HEIGHT	TIME (MIN)
100	HARRIS	JANE	O	F	60	70
101	HILL	GLENN	Y	M	65	70
102	GARCIA	PEDRO	M	M	76	85
103	HILBERT	DORIS	M	F	64	90
104	DOAKS	SALLY	Y	F	62	94
105	JONES	SUE	Y	F	64	95
106	SMITH	PETE	M	M	73	100
107	DOE	JANE	O	F	66	100
108	BRADY	PETE	O	M	73	100
109	BRADY	JOE	O	M	71	120
110	HEEBER	SALLY	M	F	59	125
111	DOLTZ	HAL	O	M	76	130
112	PEEBLES	AL	Y	M	76	63

FIGURE E-18   Runner data used to illustrate built-in functions

The data is the same as that shown in Figure E-1, except that Figure E-18 includes a column for the runners' height in inches.

MIN and MAX Functions

The MIN function determines the smallest value in a range of data. The MAX function returns the largest. Say that we want to know the fastest time for all runners, which would be the minimum time in column G. The MIN function computes the smallest value in a set of values. The set of values could be a data range, or it could be a series of cell addresses separated by commas. The syntax of the MIN function is as follows:

- MIN(set of data)

To show the minimum time in cell C16, you would enter the formula shown in the formula bar in Figure E-19:

	C16	▼	*fx*	=MIN(G2:G14)		
	A	B	C	D	E	F	G
1	RUNNER#	LAST	FIRST	AGE	GENDER	HEIGHT	TIME (MIN)
2	100	HARRIS	JANE	O	F	60	70
3	101	HILL	GLENN	Y	M	65	70
4	102	GARCIA	PEDRO	M	M	76	85
5	103	HILBERT	DORIS	M	F	64	90
6	104	DOAKS	SALLY	Y	F	62	94
7	105	JONES	SUE	Y	F	64	95
8	106	SMITH	PETE	M	M	73	100
9	107	DOE	JANE	O	F	66	100
10	108	BRADY	PETE	O	M	73	100
11	109	BRADY	JOE	O	M	71	120
12	110	HEEBER	SALLY	M	F	59	125
13	111	DOLTZ	HAL	O	M	76	130
14	112	PEEBLES	AL	Y	M	76	63
15							
16	MINIMUM TIME:		63				

FIGURE E-19   MIN function in cell C16

(Assume that you typed the label "MINIMUM TIME:" into cell A16.) You can see that the fastest time is 63 minutes.

To see the slowest time in cell G16, use the MAX function, whose syntax parallels that of the MIN function, except that the largest value in the set is determined. See Figure E-20.

	G16	▼	*fx*	=MAX(G2:G14)		
	A	B	C	D	E	F	G
1	RUNNER#	LAST	FIRST	AGE	GENDER	HEIGHT	TIME (MIN)
2	100	HARRIS	JANE	O	F	60	70
3	101	HILL	GLENN	Y	M	65	70
4	102	GARCIA	PEDRO	M	M	76	85
5	103	HILBERT	DORIS	M	F	64	90
6	104	DOAKS	SALLY	Y	F	62	94
7	105	JONES	SUE	Y	F	64	95
8	106	SMITH	PETE	M	M	73	100
9	107	DOE	JANE	O	F	66	100
10	108	BRADY	PETE	O	M	73	100
11	109	BRADY	JOE	O	M	71	120
12	110	HEEBER	SALLY	M	F	59	125
13	111	DOLTZ	HAL	O	M	76	130
14	112	PEEBLES	AL	Y	M	76	63
15							
16	MINIMUM TIME:		63		MAXIMUM TIME:		130

FIGURE E-20   MAX function in cell G16

AVERAGE, ROUND, and ROUNDUP Functions

The AVERAGE function computes the average of a set of values. Figure E-21 shows the use of the AVERAGE function in cell C17:

C17		fx	=AVERAGE(G2:G14)			
	A	B	C	D	E	F	G
1	RUNNER#	LAST	FIRST	AGE	GENDER	HEIGHT	TIME (MIN)
2	100	HARRIS	JANE	O	F	60	70
3	101	HILL	GLENN	Y	M	65	70
4	102	GARCIA	PEDRO	M	M	76	85
5	103	HILBERT	DORIS	M	F	64	90
6	104	DOAKS	SALLY	Y	F	62	94
7	105	JONES	SUE	Y	F	64	95
8	106	SMITH	PETE	M	M	73	100
9	107	DOE	JANE	O	F	66	100
10	108	BRADY	PETE	O	M	73	100
11	109	BRADY	JOE	O	M	71	120
12	110	HEEBER	SALLY	M	F	59	125
13	111	DOLTZ	HAL	O	M	76	130
14	112	PEEBLES	AL	Y	M	76	63
15							
16	MINIMUM TIME:		63		MAXIMUM TIME:		130
17	AVERAGE TIME:		95.53846				

FIGURE E-21 AVERAGE function in cell C17

Notice that the value shown is a real number with many digits. What if you wanted to have the value rounded to a certain number of digits? Of course, you could format the output cell, but doing that only changes what is shown on the screen. You want the cell's contents actually to *be* the rounded number. Therefore, you need to use the ROUND function. Its syntax is:

- ROUND(number, number of digits)

Figure E-22 shows the rounded average time (with two decimal places) in cell G17.

G17		fx	=ROUND(C17,2)			
	A	B	C	D	E	F	G
1	RUNNER#	LAST	FIRST	AGE	GENDER	HEIGHT	TIME (MIN)
2	100	HARRIS	JANE	O	F	60	70
3	101	HILL	GLENN	Y	M	65	70
4	102	GARCIA	PEDRO	M	M	76	85
5	103	HILBERT	DORIS	M	F	64	90
6	104	DOAKS	SALLY	Y	F	62	94
7	105	JONES	SUE	Y	F	64	95
8	106	SMITH	PETE	M	M	73	100
9	107	DOE	JANE	O	F	66	100
10	108	BRADY	PETE	O	M	73	100
11	109	BRADY	JOE	O	M	71	120
12	110	HEEBER	SALLY	M	F	59	125
13	111	DOLTZ	HAL	O	M	76	130
14	112	PEEBLES	AL	Y	M	76	63
15							
16	MINIMUM TIME:		63		MAXIMUM TIME:		130
17	AVERAGE TIME:		95.53846		ROUNDED AVERAGE		95.54

FIGURE E-22 ROUND function used in cell G17

To achieve this output, cell C17 was used as the value to be rounded. Recall from Figure E-21 that cell C17 had the formula =AVERAGE(G2:G14). The following ROUND formula would produce the same output in cell G17: =ROUND(AVERAGE(G2:G14),2). In this case, Excel evaluates the formula "inside out." First, the AVERAGE function is evaluated, yielding the average with many digits. That value is then input to the ROUND function and rounded to two decimal places.

The ROUNDUP function works much like the ROUND function. ROUNDUP's output is always rounded up to the next value. For example, the value 4 would appear in a cell that contained the following formula: =ROUNDUP(3.12,0). In Figure E-22, if the formula in cell G17 had been =ROUNDUP(AVERAGE(G2:G14),2), the value 96 would have been the result. In other words, 95.54 rounded up with no decimal places becomes 96.

COUNTIF Function

The COUNTIF function counts the number of values in a range that meet a specified condition. The syntax is:

- COUNTIF(range of data, condition)

The condition is a logical expression such as "=1", ">6", or "=F". The condition is shown with quotation marks, even if a number is involved.

Assume that you want to see the number of female runners in cell C18. Figure E-23 shows the formula used.

			C18		▼		f_x	=COUNTIF(E2:E14,"F")	
	A	B	C	D	E	F	G	
1	RUNNER#	LAST	FIRST	AGE	GENDER	HEIGHT	TIME (MIN)	
2	100	HARRIS	JANE	O	F	60	70	
3	101	HILL	GLENN	Y	M	65	70	
4	102	GARCIA	PEDRO	M	M	76	85	
5	103	HILBERT	DORIS	M	F	64	90	
6	104	DOAKS	SALLY	Y	F	62	94	
7	105	JONES	SUE	Y	F	64	95	
8	106	SMITH	PETE	M	M	73	100	
9	107	DOE	JANE	O	F	66	100	
10	108	BRADY	PETE	O	M	73	100	
11	109	BRADY	JOE	O	M	71	120	
12	110	HEEBER	SALLY	M	F	59	125	
13	111	DOLTZ	HAL	O	M	76	130	
14	112	PEEBLES	AL	Y	M	76	63	
15								
16	MINIMUM TIME:		63		MAXIMUM TIME:		130	
17	AVERAGE TIME:		95.53846		ROUNDED AVERAGE:		95.54	
18	NUMBER OF FEMALES:		6					

FIGURE E-23   COUNTIF function used in cell C18

The logic of the formula is: Count the number of times that "F" appears in the data range E2:E14.

As another example of using COUNTIF, assume that column H shows the rounded ratio of the runner's height in inches to the runner's time in minutes (see Figure E-24).

	H2	▾		f_x	=ROUND(G2/F2,2)			
	A	B	C	D	E	F	G	H
1	**RUNNER#**	**LAST**	**FIRST**	**AGE**	**GENDER**	**HEIGHT**	**TIME (MIN)**	**RATIO**
2	100	HARRIS	JANE	O	F	60	70	1.17
3	101	HILL	GLENN	Y	M	65	70	1.08
4	102	GARCIA	PEDRO	M	M	76	85	1.12
5	103	HILBERT	DORIS	M	F	64	90	1.41
6	104	DOAKS	SALLY	Y	F	62	94	1.52
7	105	JONES	SUE	Y	F	64	95	1.48
8	106	SMITH	PETE	M	M	73	100	1.37
9	107	DOE	JANE	O	F	66	100	1.52
10	108	BRADY	PETE	O	M	73	100	1.37
11	109	BRADY	JOE	O	M	71	120	1.69
12	110	HEEBER	SALLY	M	F	59	125	2.12
13	111	DOLTZ	HAL	O	M	76	130	1.71
14	112	PEEBLES	AL	Y	M	76	63	0.83
15								
16	MINIMUM TIME:		63		MAXIMUM TIME:		130	
17	AVERAGE TIME:		95.53846		ROUNDED AVERAGE:		95.54	
18	NUMBER OF FEMALES:		6					

FIGURE E-24   Ratio of height to time in column H

Assume that all runners whose height in inches is less than their time in minutes will get an award. How many awards are needed? If the ratio is less than 1, an award is warranted. The COUNTIF function in cell G18 computes a count of ratios less than 1, as shown in Figure E-25.

	G18	▾		f_x	=COUNTIF(H2:H14,"<1")			
	A	B	C	D	E	F	G	H
1	**RUNNER#**	**LAST**	**FIRST**	**AGE**	**GENDER**	**HEIGHT**	**TIME (MIN)**	**RATIO**
2	100	HARRIS	JANE	O	F	60	70	1.17
3	101	HILL	GLENN	Y	M	65	70	1.08
4	102	GARCIA	PEDRO	M	M	76	85	1.12
5	103	HILBERT	DORIS	M	F	64	90	1.41
6	104	DOAKS	SALLY	Y	F	62	94	1.52
7	105	JONES	SUE	Y	F	64	95	1.48
8	106	SMITH	PETE	M	M	73	100	1.37
9	107	DOE	JANE	O	F	66	100	1.52
10	108	BRADY	PETE	O	M	73	100	1.37
11	109	BRADY	JOE	O	M	71	120	1.69
12	110	HEEBER	SALLY	M	F	59	125	2.12
13	111	DOLTZ	HAL	O	M	76	130	1.71
14	112	PEEBLES	AL	Y	M	76	63	0.83
15								
16	MINIMUM TIME:		63		MAXIMUM TIME:		130	
17	AVERAGE TIME:		95.53846		ROUNDED AVERAGE:		95.54	
18	NUMBER OF FEMALES:		6		RATIOS<1:		1	

FIGURE E-25   COUNTIF function used in cell G18

RANDBETWEEN Function

If you wanted a cell to contain a randomly generated integer in the range from 1 to 9, you would use the formula =RANDBETWEEN(1,9). Any value between 1 and 9 inclusive would be output by the formula. An example is shown in Figure E-26.

A2		f_x	=RANDBETWEEN(1,9)		
A	B	C	D	E	
1	Position				
2	9				

FIGURE E-26   RANDBETWEEN function used in cell A2

Assume that you copied and pasted the formula to generate a column of 100 numbers between 1 and 9. Every time a value was changed in the spreadsheet, Excel would recalculate the 100 RANDBETWEEN formulas to change the 100 random values. Therefore, you might want to settle on the random values once they are generated. To do this, copy the 100 values, select Paste Special, and then select Values to put the values in the same range. The contents of the cells will change from formulas to literal values.

PART 7

PRESENTATION SKILLS

GIVING AN ORAL PRESENTATION

Giving an oral presentation provides you the opportunity to practice the presentation skills you will need in the workplace. The presentations you create for the cases in this textbook will be similar to professional business presentations. You will be expected to present objective, technical results to your organization's stakeholders, and you will have to support your presentation with visual aids commonly used in the business world. During your presentation, your instructor might assign your classmates to role-play an audience of business managers, bankers, or employees, and ask them to provide feedback on your presentation.

Follow these four steps to create an effective presentation:

1. Plan your presentation.
2. Draft your presentation.
3. Create graphics and other visual aids.
4. Practice delivering your presentation.

PLANNING YOUR PRESENTATION

When planning an oral presentation, you need to be aware of your time limits, establish your purpose, analyze your audience, and gather information. This section explores each of these elements.

Knowing Your Time Limits

You need to consider your time limits on two levels. First, consider how much time you will have to deliver your presentation. For example, what are the key points in your material that can be covered in 10 minutes? The element of time is the primary constraint of any presentation. It limits the breadth and depth of your talk, and the number of visual aids that you can use. Second, consider how much time you will need for the process of preparing your presentation—drafting your presentation, creating graphics, and practicing your delivery.

Establishing Your Purpose

After considering your time limits, you must define your purpose: what you need to say and to whom you will say it. For the Access cases in this book, your purpose will be to inform and explain. For instance, a business's owners, managers, and employees may need to know how the company's database is organized and how they can use it to fill in input forms and create reports. In contrast, for the Excel cases, your purpose will be to recommend a course of action based on the results of your business model. You will make the recommendations to business owners, managers, and bankers based on the results of inputting and running various scenarios.

Analyzing Your Audience

Once you have established the purpose of your presentation, you should analyze your audience. Ask yourself: What does my audience already know about the subject? What do the audience members want to know? What do they need to know? Do they have any biases or personal agendas that I should consider? What level of technical detail is best suited to their level of knowledge and interest?

In some Access cases, you will make a presentation to an audience that might not be familiar with Access or with databases in general. In other cases, you might be giving your presentation to a business owner who started to work on a database but was not able to finish it. Tailor the presentation to suit your audience.

For the Excel cases, you are probably interpreting results for an audience of bankers or business managers. In those instances, the audience will not need to know the detailed technical aspects of how you generated your results. But what if your audience consists of engineers or scientists? They will certainly be more interested in the structure and rationale of your decision models. Regardless of the audience, your listeners need to know what assumptions you made prior to developing your spreadsheets because those assumptions might affect their opinion of your results.

Gathering Information

Because you will have just completed a case as you begin preparing your oral presentation, you will already have the basic information you need. For the Access cases, you should review the main points of the case and your goals. Make sure you include all of the points you think are important for the audience to understand. In addition, you might want to go beyond the requirements and explain additional ways in which the database could be used to benefit the organization, now or in the future.

For the Excel cases, you can refer to the tutorials for assistance in interpreting the results from your spreadsheet analysis. For some cases, you might want to use the Internet or the library to research business trends or background information that you can use to support your presentation.

DRAFTING YOUR REPORT AND PRESENTATION

When you have completed the planning stage, you are ready to begin drafting the presentation. At this point, you might be tempted to write your presentation and then memorize it word for word. Even if you could memorize your presentation verbatim, however, your delivery would sound unnatural because people use a simpler vocabulary and shorter sentences when they speak than when they write. For example, read the previous paragraph out loud as if you were presenting it to an audience.

In many business situations, you will be required both to submit a written report of your work and give a PowerPoint presentation. First write your report, and then design your PowerPoint slides as a "brief" of that report to discuss its main points. When drafting your report and the accompanying PowerPoint slides, follow this sequence:

1. Write the main body of your report.
2. Write the introduction to your report.
3. Write the conclusion to your report.
4. Prepare your presentation (the PowerPoint slides) using your report's main points.

Writing the Main Body

When you draft your report, write the body first. If you try to write the opening paragraph first, you might spend an inordinate amount of time attempting to craft your words perfectly, only to revise the introduction after you write the body of the report.

Keeping Your Audience in Mind

To write the main body, review your purpose and your audience profile. What are the main points you need to make? What are your audience's needs, interests, and technical expertise? It is important to include some basic technical details in your report and presentation, but keep in mind the technical expertise of your audience.

Remember that the people reading your report or listening to your presentation have their own agendas—put yourself in their places and ask "What do I need to get out of this presentation?" For example, in the Access cases, an employee might want to know how to enter information on a form, but the business owner might be more interested in learning how to generate queries and reports. You need to address their different needs in your presentation. For example, you might say, "And now, let's look at how data entry associates can input data into this form."

Similarly, in the Excel cases, your audience will consist of business owners, managers, bankers, and perhaps some technical professionals. The owners and managers will be concerned with profitability, growth, and customer service. In contrast, the bankers' main concern will be repayment of a loan. Technical professionals will be more concerned with how well your decision model is designed, along with the credibility of the results. You need to address the interests of each group.

Using Transitions and Repetition in Your Presentation

During your presentation, bear in mind that your audience is not reading the text of your report, so you need to include transitions to compensate. Words such as *next, first, second,* and *finally* will help the audience follow the sequence of your ideas. Words such as *however, in contrast, on the other hand,* and *similarly* will help the audience follow shifts in thought. You also can use your voice and hand gestures to convey emphasis.

Also consider using body language to emphasize what you say. For instance, if you list three items, you can use your fingers to tick off each item as you discuss it. Similarly, if you state that profits will be flat, you can make a level motion with your hand for emphasis.

You may be speaking behind a podium or standing beside a projection screen, or both. If you feel uncomfortable standing in one place and you can walk without blocking the audience's view of the screen, feel free to move around. You can emphasize a transition by changing your position. If you tend to fidget, shift, or rock from one foot to the other, try to anchor yourself. A favorite technique of some speakers is to come from behind the podium and place one hand on the side of it while speaking. They get the anchoring effect of the podium while removing the barrier it places between them and the audience. Use the stance or technique that makes you feel most comfortable, as long as your posture or actions do not distract the audience.

As you draft your presentation, repeat key points to emphasize them. For example, suppose your main point is that outsourcing labor will provide the greatest gains in net income. Begin by previewing that concept, and state that you will demonstrate how outsourcing labor will yield the biggest profits. Then provide statistics that support your claim, and show visual aids that graphically illustrate your point. Summarize by repeating your point: "As you can see, outsourcing labor does yield the biggest profits."

Relying on Graphics to Support Your Talk

As you write the main body, think of how to integrate graphics into your presentation. Do not waste words with a long description if a graphic can bring instant comprehension. For instance, instead of describing how information from a query can be turned into a report, show the query and a completed report. Figures F-1 and F-2 illustrate an Access query and the resulting report.

Order Query 1

Customer Name	City	Product Name	Qty	Price per Unit	Total
Applewood Restaurant	Martinsburg	Frozen Alligator on a Stick	20	$27.99	$559.80
Applewood Restaurant	Martinsburg	Nogales Chipotle Sauce	15	$11.49	$172.35
Applewood Restaurant	Martinsburg	Mom's Deep Dish Apple Pie	12	$12.49	$149.88
Fresh Catch Fishery	Salem	Brumley's Seafood Cocktail Sauce	24	$4.79	$114.96
Fresh Catch Fishery	Salem	NY Smoked Salmon	21	$21.99	$461.79
Fresh Catch Fishery	Salem	Mama Mia's Tiramisu	15	$17.99	$269.85
Jimmy's Crab House	Elkton	Frozen Alligator on a Stick	12	$27.99	$335.88
Jimmy's Crab House	Elkton	Brumley's Seafood Cocktail Sauce	24	$4.79	$114.96
Jimmy's Crab House	Elkton	Mama Mia's Tiramisu	18	$17.99	$323.82
Jimmy's Crab House	Elkton	Mom's Deep Dish Apple Pie	36	$12.49	$449.64

FIGURE F-1 Access query

July 2010 Orders--Fine Foods, Inc. Sunday, November 21, 2010 11:54:22 AM

Customer Name	City	Product Name	Qty	Price per Unit	Total
Applewood Restaurant	Martinsburg	Frozen Alligator on a Stick	20	$27.99	$559.80
Applewood Restaurant	Martinsburg	Nogales Chipotle Sauce	15	$11.49	$172.35
Applewood Restaurant	Martinsburg	Mom's Deep Dish Apple Pie	12	$12.49	$149.88
Fresh Catch Fishery	Salem	Brumley's Seafood Cocktail Sauce	24	$4.79	$114.96
Fresh Catch Fishery	Salem	NY Smoked Salmon	21	$21.99	$461.79
Fresh Catch Fishery	Salem	Mama Mia's Tiramisu	15	$17.99	$269.85
Jimmy's Crab House	Elkton	Frozen Alligator on a Stick	12	$27.99	$335.88
Jimmy's Crab House	Elkton	Brumley's Seafood Cocktail Sauce	24	$4.79	$114.96
Jimmy's Crab House	Elkton	Mama Mia's Tiramisu	18	$17.99	$323.82
Jimmy's Crab House	Elkton	Mom's Deep Dish Apple Pie	36	$12.49	$449.64

Total Orders $2,952.93

Page 1 of 1

FIGURE F-2 Access report

Also consider what kinds of graphic media are available and how well you can use them. Your employer will expect you to be able to use Microsoft PowerPoint to prepare your presentation as a slide show. Luckily, many college freshmen are required to take an introductory course that covers Microsoft Office and PowerPoint. If you are not familiar with PowerPoint, several excellent tutorials on the Web can help you learn the basics.

Anticipating the Unexpected

Even though you are only drafting your report and presentation at this stage, eventually you will answer questions from the audience. Being able to handle questions smoothly is the mark of a business professional. The first steps to addressing audience questions are being able to anticipate them and preparing your answers.

You will not use all the facts you gather for your report or presentation. However, as you draft your report, you might want to jot down those facts and keep them handy, in case you need them to answer questions from the audience. PowerPoint has a Notes section where you can include notes for each slide and print them to help you answer questions that arise during your presentation. You will learn how to print notes for your slides later in the tutorial.

The questions you receive depend on the nature of your presentation. For example, during a presentation of an Excel decision model, you might be asked why you are not recommending a certain course of action, or why you left it out of your report. If you have already prepared notes that anticipate such questions, you will probably remember your answers without even having to refer to the notes.

Another potential problem is determining how much technical detail you should display in your slides. In one sense, writing your report will be easier because you can include any graphics, tables, or data you want. Because you have a time limit for your presentation, the question of what to include or leave out becomes more challenging. One approach to this problem is to create more slides than you think you need, and then use the Hide Slide option in PowerPoint to "hide" the extra slides. For example, you might create slides that contain technical details you do not think you will have time to present. However, if you are asked for more details on a particular technical point, you can "unhide" a slide and display the detailed information needed to answer the question. You will learn more about the Hide Slide and Unhide Slide options later in the tutorial.

Writing the Introduction

After you have written the main body of your report and presentation, you can develop an introduction. The introduction should be only a paragraph or two, and it should preview the main points you will cover.

For some of the Access cases, you might want to include general information about databases: what they can do, why they are used, and how they can help a company become more efficient and profitable. You will not need to say much about the business operation because the audience already works for the company.

For the Excel cases, you might want to include an introduction of the general business scenario and describe any assumptions you used to create and run your decision support models. Excel is used for decision support, so you should describe the decision criteria you selected for the model.

Writing the Conclusion

Every good report or presentation needs a good ending. Do not leave the audience hanging. Your conclusion should be brief—only a paragraph or two—and it should give your presentation a sense of closure. Use the conclusion to repeat your main points or, for the Excel cases, to recap your findings and recommendations.

On many occasions, information learned during a business project reveals new opportunities for other projects. Your conclusion should provide closure for the immediate project, but if the project reveals possibilities for future improvements, include them in a Path Forward statement.

CREATING GRAPHICS

Using visual aids is a powerful means of getting your point across and making it understandable to your audience. Visual aids come in a variety of forms, some of which are more effective than others. The integrated graphics tools in Microsoft Office can help you prepare a presentation with powerful impact.

Choosing Presentation Media

The media you use will depend on the situation and the media you have available, but remember: *You must maintain control of the media or you will lose the attention of your audience.*

The following list highlights the most common media used in a classroom or business conference room, along with their strengths and weaknesses:

- **PowerPoint slides and a projection system**—These are the predominant presentation media for academic and business use. You can use a portable screen and a simple projector hooked up to a PC, or you can use a full multimedia center. Also, although they are not yet universal in business, touch-sensitive projection screens (for example, Smartboard™ technology) are gaining popularity in college classrooms. The ability to project and display slides, video and sound clips, and live Web pages makes the projection system a powerful presentation tool. *Negatives:* Depending on the complexity of the equipment, you might have difficulties setting it up and getting it to work properly. Also, you often must darken the room to use the projector, and it may be difficult to refer to written notes during your presentation. When using presentation media, you must be able to access and load your PowerPoint file easily. Make sure your file is available from at least two sources that the equipment can access, such as a thumb drive, CD, DVD, or online folder.

- **Handouts**—You can create handouts of your presentation for the audience, which once was the norm for many business meetings. Handouts allow the audience to take notes on applicable slides. If the numbers on a screen are hard to read from the back of the room, your audience can refer to their handouts. With the growing emergence of "green" business practices, however, unnecessary paper use is being discouraged. Many businesses now require reports and presentation slides to be posted at a common site where the audience can access them later. Often this site is a "public" drive on a business network. *Negatives:* Giving your audience reading material may distract their attention from your presentation. They could read your slides and possibly draw wrong conclusions from them before you have a chance to explain them.

- **Overhead transparencies**—Transparencies are rarely used anymore in business, but some academics prefer them, particularly if they have to write numbers, equations, or formulas on a display large enough for students to see from the back row in a lecture hall. *Negatives:* Transparencies require an overhead projector, and frequently their edges are visually distorted due to the design of the projector lens. You have to use special transparency sheets in a photocopier to create your slides. For both reasons, it is best to avoid using "overheads."

- **Whiteboards**—Whiteboards are common in both the business conference room and the classroom. They are useful for posting questions or brainstorming, but you should not use one in your presentation. *Negatives:* You have to face away from your audience to use a whiteboard, and if you are not used to writing on one, it can be difficult to write text that is large enough and legible. Use whiteboards only to jot down questions or ideas that you will check on after the presentation is finished.

- **Flip charts**—Flip charts (also known as easel boards) are large pads of paper on a portable stand. They are used like whiteboards, except that you do not erase your work when the page is full—you flip over to a fresh sheet. Like whiteboards, flip charts are useful for capturing questions or ideas that you want to research after the presentation is finished. Flip charts have the same negatives as white boards. Their one advantage is that you can tear off the paper and take it with you when you leave.

Creating Graphs and Charts

Strictly speaking, charts and graphs are not the same thing, although many graphs are referred to as charts. Usually charts show relationships and graphs show change. However, Excel makes no distinction, and calls both entities charts.

Charts are easy to create in Excel. Unfortunately, the process is so easy that people frequently create graphics that are meaningless or misleading, or that inaccurately reflect the data represented. This section explains how to select the most appropriate graphics.

You should use pie charts to display data that is related to a whole. For example, you might use a pie chart when breaking down manufacturing costs into Direct Materials, Direct Labor, and Manufacturing Overhead, as shown in Figure F-3. (Note that when you create a pie chart, Excel will convert the numbers you want to graph into percentages of 100.)

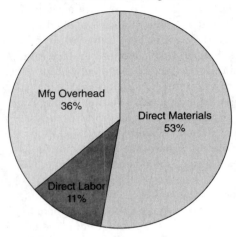

FIGURE F-3   Pie chart: appropriate use

You would *not*, however, use a pie chart to display a company's sales over a three-year period. For example, the pie chart in Figure F-4 is meaningless because it is not useful to think of the period "as a whole" or the years as its "parts."

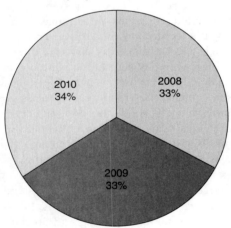

FIGURE F-4   Pie chart: inappropriate use

You should use vertical bar charts (also called column charts) to compare several amounts at the same time, or to compare the same data collected for successive periods of time. The same type of company sales data shown in Figure F-4 can be compared correctly using a vertical bar chart (see Figure F-5).

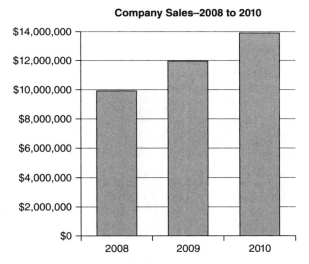

FIGURE F-5 Column chart: appropriate use

As another example, you might want to compare the sales revenues from several different products. You can use a clustered bar chart to show changes in each product's sales over time, as in Figure F-6. This type of chart is called a "clustered column" chart in Excel.

When building a chart, include labels that explain the graphics. For instance, when using a graph with an x- and y-axis, you should show what each axis represents so your audience does not puzzle over the graphic while you are speaking. Figures F-6 and F-7 illustrate the necessity of good labels.

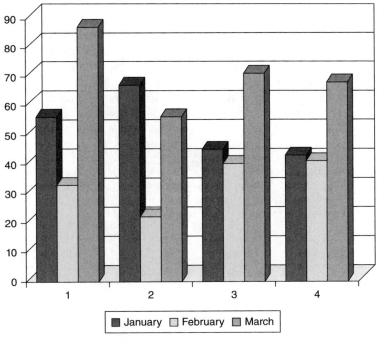

FIGURE F-6 3-D clustered column graph without labels

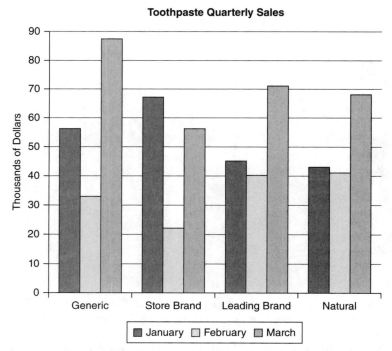

FIGURE F-7   Graph with labels

In Figure F-6, the graph has no title and neither axis is labeled. Are the amounts in units or dollars? What elements are represented by each cluster of bars? In contrast, Figure F-7 provides a comprehensive snapshot of product sales, which would support a talk rather than create confusion.

Another common pitfall of visual aids is charts that have a misleading premise. For example, suppose you want to show how sales are distributed among your inventory, and their contribution to net income. If you simply take the number of items sold in a given month, as displayed in Figure F-8, the visual fails to give your audience a sense of the actual dollar value of those sales. It is far more appropriate and informative to graph the net income for the items sold instead of the number of items sold. The graph in Figure F-9 provides a more accurate picture of which items contribute the most to net income.

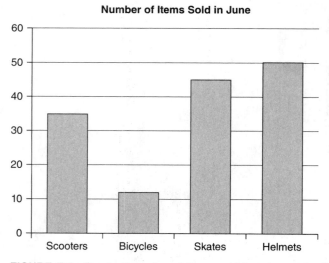

FIGURE F-8   Graph of number of items sold that does not reflect generated income

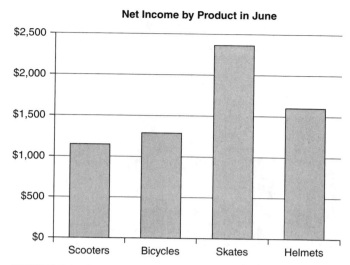

FIGURE F-9   Graph of net income by item sold

You should also avoid putting too much data in a single comparative chart. For example, assume that you want to compare monthly mortgage payments for two loan amounts with different interest rates and time frames. You have a spreadsheet that computes the payment data, as shown in Figure F-10.

	A	B	C	D	E	F	G
1	**Calculation of Monthly Payment**						
2	Rate	6.00%	6.10%	6.20%	6.30%	6.40%	6.50%
3	Amount	$ 100,000	$ 100,000	$ 100,000	$ 100,000	$ 100,000	$ 100,000
4	Payment (360 Payments)	$ 599	$ 605	$ 612	$ 618	$ 625	$ 632
5	Payment (180 Payments)	$ 843	$ 849	$ 854	$ 860	$ 865	$ 871
6	Amount	$ 150,000	$ 150,000	$ 150,000	$ 150,000	$ 150,000	$ 150,000
7	Payment (360 Payments)	$ 899	$ 908	$ 918	$ 928	$ 938	$ 948
8	Payment (180 Payments)	$ 1,265	$ 1,273	$ 1,282	$ 1,290	$ 1,298	$ 1,306

FIGURE F-10   Calculation of monthly payment

In Excel, it is possible (but not advisable) to capture all of the information in a single clustered column chart, as shown in Figure F-11.

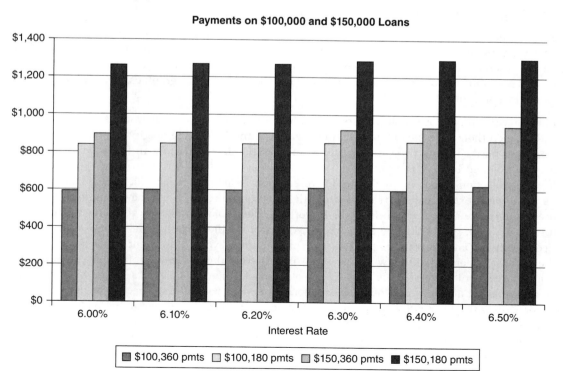

FIGURE F-11   Too much information in one chart

The chart contains a great deal of information. Putting the $100,000 loan payments and $150,000 payments in the same "cluster" may confuse the readers. They would probably find it easier to understand one chart that summarizes the $100,000 loan (see Figure F-12) and a second chart that covers the $150,000 loan.

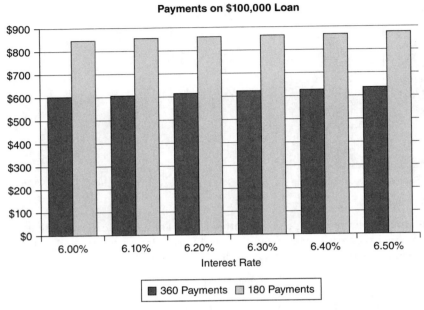

FIGURE F-12   Good balance of information

You could then augment the charts with text that summarizes the main differences between the payments for each loan amount. In that fashion, the reader is led step by step through the analysis.

Excel no longer has a Chart Wizard; instead, the Insert tab includes a Charts group. Once you create a chart and click it, three chart-specific tabs appear under a Chart Tools heading in the Ribbon to assist you with chart design, layout, and formatting. If you are unfamiliar with the charting tools in Excel, ask your instructor for guidance or refer to the many Excel tutorials on the Web.

Creating PowerPoint Presentations

PowerPoint presentations are easy to create. When you open PowerPoint, it starts a new presentation for you. You can select from many different themes, styles, and slide layouts by clicking the Design tab. If none of PowerPoint's default themes suit you, you can download theme "templates" from Microsoft Office online. When choosing a theme and style for your slides, such as background colors or graphics, fonts, and fills, keep the following guidelines in mind:

- In older versions of PowerPoint, users were advised to avoid pastel backgrounds or theme colors, and to keep their slide backgrounds dark. Because of the increasing quality of graphics in both computer hardware and projection systems, most of the default themes in PowerPoint will project well and be easy to read.
- If your projection screen is small or your presentation room is large, consider using boldface type for all of your text to make it readable from the back of the room. If you have time to visit the presentation site beforehand, bring your PowerPoint file, project a slide on the screen, and look at it from the back row of the audience area. If you can read the text, the font is large enough.
- Use transitions and animations to keep your presentation lively, but do not go overboard with them. Swirling letters and pinwheeling words can distract the audience from your presentation.
- It is an excellent idea to animate the text on your slides with entrance effects so that only one bullet point appears at a time when you click the mouse (or when you tap the screen using a touch-sensitive board). This approach prevents your audience from reading ahead of the bullet point being discussed and keeps their attention on you. Entrance effects can be incorporated and managed using the Add Animation button in PowerPoint 2010, as shown in Figures F-13 and F-14.

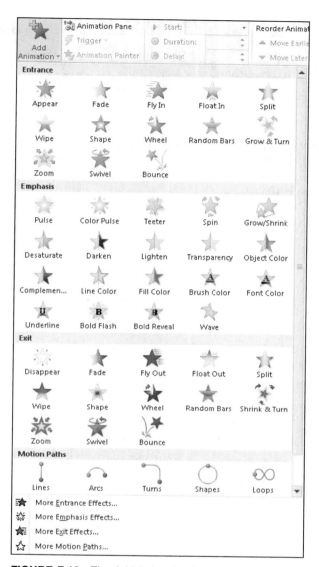

FIGURE F-13   The Add Animation button on the Ribbon in PowerPoint 2010

FIGURE F-14   Add Entrance Effect window

NOTE—DIFFERENCES IN POWERPOINT ANIMATION TOOLS—2010 VS. 2007

The structure of the animation tools has changed considerably from PowerPoint 2007 to the 2010 version. The Custom Animation button and pane are both gone. Most of the custom animation tools are now incorporated using the Add Animation button in PowerPoint 2010. The look and feel is different, but the interface is more intuitive and easier to use. You can still use an animation pane to organize and edit your animations within a slide.

- Consider creating PowerPoint slides that have a section for your notes. You can print the notes from the Print dialog box by choosing Notes Pages from the "Print what" drop-down menu, as shown in Figure F-15. Each slide will be printed at half its normal size, and your notes will appear beneath each slide, as shown in Figure F-16.

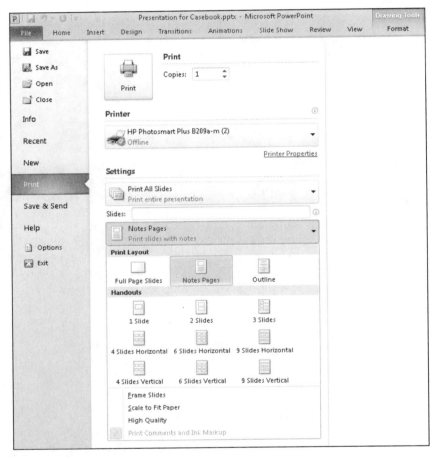

FIGURE F-15   Printing notes pages

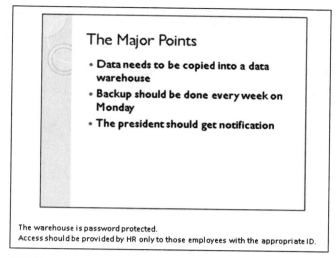

FIGURE F-16   Sample notes page

- Finally, you should check your PowerPoint slides on a projection screen before your presentation. Information that looks good on a computer display may not be readable on the projection screen.

Using Visual Aids Effectively

Make sure you choose the visual aids that will work most effectively, and that you have enough without using too many. How many is too many? The amount of time you have to speak will determine the number of visual aids you should use, as will your target audience. A good rule of thumb is to allow at least one minute to present each PowerPoint slide. Leave a minimum of two minutes for audience questions after a 10-minute presentation, and allow up to 25 percent of your total presentation time to address questions after longer presentations. (For example, for a 20-minute presentation, figure on taking five minutes for questions.) For a 10-minute talk, try to keep the body of your presentation to eight slides or less. Your target audience will also influence your selection of visual aids. For instance, your slides will need more graphics and animation if you are addressing a group of teenagers than if you are presenting to a board of directors. Remember to use visual aids to emphasize your main points, not to detract from them.

Review each of your slides and visual aids to make sure it meets the following criteria:

- The font size of the text is large enough to read from the back of the presentation area.
- The slide or visual aid is as perfect as possible, and does not contain misleading graphics, typographical errors, or misspelled words—the quality of your work is a direct reflection on you.
- The content of your visual aid is relevant to the key points of your presentation.
- The slide or visual aid does not detract from your message. Your animations, pictures, and sound effects should support the text. Your visuals should look professional.
- A visual aid should look good in the presentation environment. If possible, rehearse your PowerPoint presentation beforehand in the room where you will give the actual presentation. Make sure you can read your slides easily from the back row of seats in the room. If you have a friend who can sit in, ask her or him to listen to your voice from the back row of seats. If you have trouble projecting your voice clearly, consider using a microphone for your presentation.
- All numbers should be rounded unless decimal places or pennies are crucial. For example, your company might only pay fractions of a cent per Web hit, but this cost may become significant after millions of Web hits.
- Slides should not look too busy or crowded. Many PowerPoint experts have a "6 by 6" rule for bullet points on a slide, which means you should include no more than six bullet points per slide and no more than six words per bullet point. Also avoid putting too many labels or pictures on a slide. Clip art can be "cutesy" and therefore has no place in a professional business presentation. A well-selected picture or two can add emphasis to the theme of a slide. For examples of a slide that is too busy versus one that conveys its points succinctly, see Figures F-17 and F-18.

Major Points

- Data needs to be copied into a data warehouse
- Backup should be done every week on Monday
- The president should get notification
- The vice president should get notification
- The data should be available on the Web
- Web access should be on a secure server
- HR sets passwords
- Only certain personnel in HR can set passwords
- Users need to show ID to obtain a password
- ID cards need to be the latest version

FIGURE F-17   Busy slide

The Major Points

- Data needs to be copied into a data warehouse
- Backup should be done every week on Monday
- The president should get notification

FIGURE F-18   Slide with appropriate number of bullet points and a supporting photo

You may find that you have created more slides than you have time to present, and you are unsure of which slides you should delete. Some may have data that an audience member might ask about. Fortunately, PowerPoint lets you "hide" slides; these hidden slides will not be displayed in Slide Show view unless you "unhide" them in Normal view. Hiding slides is an excellent way to keep detailed data handy in case your audience asks to see it. Figure F-19 shows how to hide a slide in a PowerPoint presentation. Click the slide you want to hide, right-click the mouse, and then click Hide Slide in the menu to mark the slide as hidden in the presentation. To unhide the slide, click it, right-click the mouse, and click Unhide Slide from the menu. Click the slide to display it in Slide Show view.

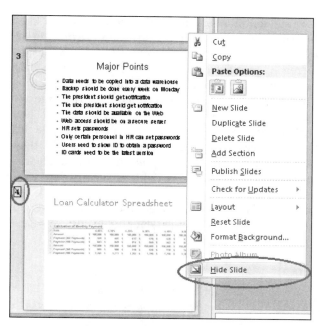

FIGURE F-19 Hiding a slide in PowerPoint

PRACTICING YOUR DELIVERY

Surveys indicate that public speaking is the greatest fear of most people. However, fear or nervousness can be channeled into positive energy to do a good job. Remember that an audience is not likely to think you are nervous unless you fidget or your voice cracks. Audience members want to hear what you have to say, so think about them and their interests—not about how you feel.

Your presentations for the cases in this textbook will occur in a classroom setting with 20 to 40 students. Ask yourself: Am I afraid when I talk to just one or two of my classmates? The answer is probably no. In addition, they will all have to give presentations as well. Think of your presentation as an extended conversation with several classmates. Let your gaze move from person to person, making brief eye contact with each of them randomly as you speak. As your focus moves from one person to another, think to yourself: I am speaking to one person at a time. As you become more proficient in speaking before a group, your gaze will move naturally among audience members.

Tips for Practicing Your Delivery

Giving an effective presentation is not the same as reading a report to an audience. You should rehearse your message well enough so that you can present it naturally and confidently, with your slides or other visual aids smoothly intermingled with your speaking. The following tips will help you hone the effectiveness of your delivery:

- Practice your presentation several times, and use your visual aids when you practice.
- Show your slides at the right time. Luckily, PowerPoint makes this easy; you can click the slide when you are ready to talk about it. Use cues as necessary in your speaker's notes.
- Maintain eye and voice contact with the audience when using the visual aid. Do not turn your back on your audience. It is acceptable to turn sideways to glance at your slide. A popular trick of experienced speakers is to walk around and steal a glance at the slide while they are moving.
- Refer to your visual aids in your talk, and use hand gestures where appropriate. Do not ignore your own visual aid, but do not read it to your audience—they can read for themselves.
- Keep in mind that your slides or visual aids should support your presentation, not *be* the presentation. Do not try to crowd the slide with everything you plan to say. Use the slides to illustrate key points and statistics, and fill in the rest of the content with your talk.
- Check your time, especially when practicing. If you stay within the time limit when practicing, you will probably finish a minute or two early when you actually give the presentation. You will be a little nervous and will talk a little faster to a live audience.

- Use numbers effectively. When speaking, use rounded numbers; otherwise, you will sound like a computer. Also make numbers as meaningful as possible. For example, instead of saying "in 83 percent of cases," say "in five out of six cases."
- Do not extrapolate, speculate, or otherwise "reach" to interpret the output of statistical models. For example, suppose your Excel model has many input variables. You might be able to point out a trend, but often you cannot say with mathematical certainty that if a company employs the inputs in the same combination, it will get the same results.
- Some people prefer recording their presentation and playing it back to evaluate themselves. It is amazing how many people are shocked when they hear their recorded voice—and usually they are not pleased with it. In addition, you will hear every *um, uh, well, you know*, throat-clearing noise, and other verbal distraction in your speech. If you want impartial feedback on your presentation, have a friend listen to it.
- If you use a pointer, be careful where you wave it. It is not a light saber, and you are not Luke Skywalker. Unless you absolutely have to use one to point out crucial data on a slide, leave the pointer home.

Handling Questions

Fielding questions from an audience can be tricky because you cannot anticipate all of the questions you might be asked. When answering questions from an audience, *treat everyone with courtesy and respect.* Use the following strategies to handle questions:

- Try to anticipate as many questions as possible, and prepare answers in advance. Remember that you can gather much of the information to prepare these answers while drafting your presentation. The Notes section under each slide in PowerPoint can be a good place to enter anticipated questions and your answers. Hidden slides can also contain the data you need to answer questions about important details.
- Mention at the beginning of your talk that you will take questions at the end of the presentation, which helps prevent questions from interrupting the flow and timing of your talk. In fact, many PowerPoint presentations end with a Questions slide. If someone tries to interrupt, say that you will be happy to answer the question when you are finished, or that the next graphic answers the question. Of course, this point does not apply to the company CEO—you *always* stop to answer the CEO's questions.
- When answering a question, a good practice is to repeat the question if you have any doubt that the entire audience heard it. Then deliver your answer to the whole audience, but make sure you close by looking directly at the person who asked the question.
- Strive to be informative, not persuasive. In other words, use facts to answer questions. For instance, if someone asks your opinion about a given outcome, you might show an Excel slide that displays the Solver's output; then you can use the data as the basis for answering the question. In that light, it is probably a good idea to have computer access to your Excel model or Access database if your presentation venue permits it, but avoid using either unless you absolutely need it.
- If you do not know the answer to a question, it is acceptable to say so, and it is certainly better than trying to fake the answer. For instance, if someone asks you the difference between the Simplex LP and GRG solving methods in Excel Solver, you might say, "That is an excellent question, but I really don't know the answer—let me research it and get back to you." Then follow up after the presentation by researching the answer and contacting the person who asked the question.
- Signal when you are finished. You might say that you have time for one more question. Wrap up the talk yourself and thank your audience for their attention.

Handling a "Problem" Audience

A "problem" audience or a heckler is every speaker's nightmare. Fortunately, this experience is rare in the classroom: Your audience will consist of classmates who also have to give presentations, and your instructor will be present to intervene in case of problems.

Heckling can be a common occurrence in the political arena, but it does not happen often in the business world. Most senior managers will not tolerate unprofessional conduct in a business meeting. However, fellow business associates might challenge you in what you perceive as a hostile manner. If so, remain calm, be professional, and rely on facts. The rest of the audience will watch to see how you react—if you behave professionally, you make the heckler appear unprofessional by comparison and gain the empathy of the audience.

A more common problem is a question from an audience member who lacks technical expertise. For instance, suppose you explained how to enter data into an Access form, but someone did not understand your explanation. Ask the questioner what part of the explanation was confusing. If you can answer the question briefly and clearly, do so. If your answer turns into a time-consuming dialogue, offer to give the person a one-on-one explanation after the presentation.

Another common problem is receiving a question that you have already answered. The best solution is to give the answer again, as briefly as possible, using different words in case your original answer confused the person. If someone persists in asking questions that have obvious answers, you might ask the audience, "Who would like to answer that question?" The questioner should get the hint.

PRESENTATION TOOLKIT

You can use the form in Figure F-20 for preparation, the form in Figure F-21 for evaluation of Access presentations, and the form in Figure F-22 for evaluation of Excel presentations.

Preparation Checklist

Facilities and Equipment

☐ The room contains the equipment that I need.
☐ The equipment works and I've tested it with my visual aids.
☐ Outlets and electrical cords are available and sufficient.
☐ All the chairs are aligned so that everyone can see me and hear me.
☐ Everyone will be able to see my visual aids.
☐ The lights can be dimmed when/if needed.
☐ Sufficient light will be available so I can read my notes when the lights are dimmed.

Presentation Materials

☐ My notes are available, and I can read them while standing up.
☐ My visual aids are assembled in the order that I'll use them.
☐ A laser pointer or a wand will be available if needed.

Self

☐ I've practiced my delivery.
☐ I am comfortable with my presentation and visual aids.
☐ I am prepared to answer questions.
☐ I can dress appropriately for the situation.

FIGURE F-20   Preparation checklist

Evaluating Access Presentations

Course: _____   **Speaker:** _____   **Date:** _____

Rate the presentation by these criteria:
4= Outstanding   3= Good   2= Adequate   1= Needs Improvement
N/A= Not Applicable

Content

_____   The presentation contained a brief and effective introduction.

_____   Main ideas were easy to follow and understand.

_____   Explanation of database design was clear and logical.

_____   Explanation of using the form was easy to understand.

_____   Explanation of running the queries and their output was clear.

_____   Explanation of the report was clear, logical, and useful.

_____   Additional recommendations for database use were helpful.

_____   Visuals were appropriate for the audience and the task.

_____   Visuals were understandable, visible, and correct.

_____   The conclusion was satisfying and gave a sense of closure.

Delivery

_____   Was poised, confident, and in control of the audience

_____   Made eye contact

_____   Spoke clearly, distinctly, and naturally

_____   Avoided using slang and poor grammar

_____   Avoided distracting mannerisms

_____   Employed natural gestures

_____   Used visual aids with ease

_____   Was courteous and professional when answering questions

_____   Did not exceed time limit

Submitted by: _____

FIGURE F-21   Form for evaluation of Access presentations

Evaluating Excel Presentations

Course: _____ Speaker: _____ Date: _____

Rate the presentation by these criteria:
4=Outstanding   3=Good   2=Adequate   1=Needs Improvement
N/A=Not Applicable

Content

_____ The presentation contained a brief and effective introduction.

_____ The explanation of assumptions and goals was clear and logical.

_____ The explanation of software output was logically organized.

_____ The explanation of software output was thorough.

_____ Effective transitions linked main ideas.

_____ Solid facts supported final recommendations.

_____ Visuals were appropriate for the audience and the task.

_____ Visuals were understandable, visible, and correct.

_____ The conclusion was satisfying and gave a sense of closure.

Delivery

_____ Was poised, confident, and in control of the audience

_____ Made eye contact

_____ Spoke clearly, distinctly, and naturally

_____ Avoided using slang and poor grammar

_____ Avoided distracting mannerisms

_____ Employed natural gestures

_____ Used visual aids with ease

_____ Was courteous and professional when answering questions

_____ Did not exceed time limit

Submitted by: _____

FIGURE F-22   Form for evaluation of Excel presentations

INDEX